DEEP TECH

DEEP TECH

DEMYSTIFYING THE BREAKTHROUGH
TECHNOLOGIES THAT WILL
REVOLUTIONIZE EVERYTHING

ERIC REDMOND

DEEP TECH
Demystifying the Breakthrough Technologies
That Will Revolutionize Everything

ISBN   978-1-5445-1895-4 *Hardcover*
        978-1-5445-1894-7 *Paperback*
        978-1-5445-1893-0 *Ebook*
        978-1-5445-1896-1 *Audiobook*

To my family, especially my kids, who sacrificed too many daddy playdates while I wrote this book—but on balance, you did get to test VR games and 3D print your own toys.

CONTENTS

0

INTRODUCTION

Imagine, for a moment, that it's December 31—2030.

You wake up a bit older and achier than you were a decade ago, but feeling pretty well all things considered. The lights in your room have illuminated slowly to mimic the sunrise, and emit a surround sound of chirping birds. It must be a Monday. Before rolling from the bed, you grab your augmented reality (AR) Apple Smart Glasses, and the first thing you see is your daily calendar hovering about eight feet from your head. For a bit of fun, the words "Happy New Year's Eve!" (brought to you by TikTok) pop up, and virtual confetti falls all around. You notice you have two NYE party invites. One corporate event and one social event suggested by your virtual AI assistant, Amy. You think a bit, then tell Amy you'll attend the social party. She responds, "Sure thing," and accepts the invite on your behalf.

As you roll out of bed, you see your sleep stats projected—data compliments of your sleep tracking smart ring. It sends the stats to your personal medical blockchain, accessible by your healthcare provider. You never had much trouble sleeping, but you get a discount on your health insurance for wearing it, and like many people, you kind of enjoy collecting statistics on yourself. That same motivation is what drives you to step on the scale, which notices you have gained a few ounces since yesterday and adjusts your calorie goals for the day. Your smart glasses respond by suggesting oatmeal, a personalized choice based on genomic markers, and it's a food you have enjoyed previously. You click an OK button hovering above a virtual steaming bowl in the air.

When you head into the kitchen, your 3D food printer extruded the ingredients for oatmeal into a bowl with a bit of brown sugar. It's pretty good, but you slice a banana in for taste, which your glasses take note of and adjust your daily calorie allotment. For the past three years, you've been letting this combination of AI and wearables suggest many of your meals, and you've maintained excellent health and weight because of it—that and the junk food ad blocker you installed.

After breakfast, you do a bit of work. While you sometimes do work with your AR glasses on, you prefer the more immersive experience of VR, so you don your Oculus Work goggles. You're immediately transported into a virtual ski lodge, where you pull up a virtual screen. You groan as you check your email (which, yes, still exists) and ask AI Amy to compose a few responses on your behalf. You give her the gist of what you want to say, and she fills out the rest, corrects for grammar, and if need be, translates the message into the language of the recipient. You also have a few VR meetings where your coworkers lounge around the lodge. As the day wanes on, your mind starts to wander to the party this evening. You've had a good day at work and think you might rather attend the corporate event instead.

You tell Amy about this change of heart, and she reminds you that you don't have any clean dress wear for a professional party, and it's too late for a same-day laundry pickup. You decide to buy something new instead. She suggests a set of professional evening wear that you might like, in your size, filtered by those which could be available before your car arrives for the party. You choose a few outfits and remove your VR headset. You stand in front of the smart mirror, and each outfit is virtually projected onto your reflection as though you were actually wearing it. After a few tries, you find a blue one you like best, and the Nordstrom app suggests a belt and shoes to go along with it. You agree, and the order is placed. Parts of the outfit are available at a warehouse within drone distance, while others will be knit on demand by a knitting machine shop one town over. However, the entire outfit will be collected by drone and AV logistics at a Nordstrom mini node and delivered to your house as a single package.

In the meantime, you shower, and the smart mirror guides you through some grooming tips, including helping you trim your hair just a bit, which has been a common practice ever since the Great Pandemic of 2020. As you relax in your robe, you hear the doorbell ring, and the outfit is neatly and minimally packaged at your front door. You change into your new clothes, grab your Apple Glasses, and run out the door to your autonomous vehicle that Amy has ordered to take you to the party. Sensing that you have left the house, the doors lock behind you.

As far as parties go, it's grand, complete with real humans mixing drinks (rather than the more common barbots). Cheers! Prost! L'chaim! As a glittery ball drops in the middle of the party room (it's basically just a drone inside a piñata), you think about the past, about the opportunities that had abounded in the previous decade. You kick yourself for not seeing some of the early signs and seizing advantage. It all seems so obvious in retrospect. But really, it's been a wild ride, from the scale out of AI and IoT (Internet of Things) to extended reality and autonomous vehicles—even 3D printed food! Who knew?

THE COMPANY OF TOMORROW

The science-fiction scenario we've described above may be long, but it's important to imagine the kind of world we could find ourselves in 2030. We're about a decade away from these devices and changes fully integrating themselves into our daily lives, but the beginnings of these revolutions are already here—they're happening around you right now. These large technology movements include AI, virtual reality, quantum computing, and much more. They will transform how we function in the next decade: manufacturing will change, along with agriculture, logistics, finance, medicine, work, life—everything. This book is for entrepreneurs who live in a world increasingly dominated by technology and who recognize that this is where the opportunities lie.

Think of it this way. Have you ever caught yourself wondering why some business people have all the luck? You know who I mean: that person who always happens to be at the right place at the right time—making the next discovery, making deal after deal, creatively innovating and solving problems and changing the world. Of course, if we dig deeper, we see it's not luck at all. These people just have a sense of what's coming and position themselves accordingly. While most of us are focused on what's happening next year, next quarter, or right now, these people are absorbing ancillary information and spotting trends that are years in the making.

But to do that, these people need an in-depth understanding

of the landscape they're living in so they can imagine what is possible. This kind of insight is not easily won, and there's no shortcut. You can't just wake up one day and decide to compete. You need to study the existing technology and grasp it at a fundamental level. You need to know it inside out. You need to imagine what it can become.

This is no easy task. Believe me, I know—this stuff can sometimes boggle my mind, and my job is cataloging deep technology. Do you know the important differences between artificial intelligence, machine learning, and neural networks? Do you know that the use cases for extended realities like augmented and virtual are far different than they may appear? If you think the cryptocurrency era of Bitcoin is over, you should see why people are still betting big on blockchain. Is there a difference between smart cities and an Apple Watch? What's the merit behind the view of autonomous vehicles that separate what Gartner and venture capitalists believe? Is 3D just a hobby for geeks, or are we a mere half decade away from fully functional 3D printed organs? Or consider the quantum computer. Should you invest in D-Wave or PsiQuantum? It depends. Do you need help optimizing your supply chain or cracking encryption schemes? Beyond specific technologies, the endless convergence of deep technologies can make your head spin.

So, many entrepreneurs give up. And in the business world, giving up means we outsource our understanding to others, who we pay to have this skill on our behalf. We tell ourselves we're in the business of ideas—someone else can handle the tech. And that's fine, to an extent. We all rely on experts. But to be the company of tomorrow, you can't run your entire digital transformation strategy on the word of pundits. When it comes to making the transformative changes in the world we all dream of, you need to know what you're doing.

There are many fine books focused on selling the importance of adopting technology for the future (we'll list some at the end of Chapter 9), but this book is not a treatise on innovation. If you don't already believe in the power of innovation to change the world, please return this book for a refund. What this book will offer is an in-depth look at those seven technologies so you can begin to understand them and their transformative possibilities—the ones that are becoming instrumental in shaping the future.

1. Artificial Intelligence
2. Extended Reality
3. Blockchain
4. Internet of Things
5. Autonomous Vehicles
6. 3D Printing
7. Quantum Computing

More about this in Chapter 1.

The CEO of the near future requires a minimum literacy in technology, just as all CEOs today must know the basics of finance and sales. In fact, if you take the time to understand the current tech landscape, you'll be far ahead of the pack. Some of this tech is already changing the game, while some will just come of age at the bottom of the decade.

When it comes to tech revolutions, I've been lucky. Like Forrest Gump, I've had a front-row seat to several revolutions over the past three decades, from the early web to quantum computation. I came of age at the birth of the web and held multiple careers around agile practices, mobile apps, virtual reality and augmented reality, big data and artificial intelligence—in addition to being a researcher, author, and educator. Over the past several years, I've directed technology, innovation, and research at a massive global sportswear brand. Through this work I've been able to connect even deeper with cutting-edge research in academia, enterprise partners, startups, governments, and of course, we invent quite a few things ourselves. From this vantage, I've seen what deep tech looks like at scale, worked with world-class organizations from MIT and Microsoft Labs to the World Economic Forum, and experimented with deep tech from autonomous vehicles to quantum computers. But beyond technology, my focus has been on tying together the threads of deep tech for practical uses and guiding business leaders through the current reality, which is this: understanding and building a deep tech strategy is the single greatest competitive opportunity of this decade.

This the central conceit of this book. The future belongs to those leaders who can bridge worlds—between the opportunities of today and the tech that will be. The opportunities of the near future will be led by subject-matter experts who are also fluent in the possibilities that technology can bring to their industries—something that AOL co-founder Steve Case calls the Third Wave. If you can make the transition, if you can embrace deep tech, the company of tomorrow could be yours.

1

WHAT IS DEEP TECH?

"What seems natural to us is probably just something familiar in a long tradition that has forgotten the unfamiliar source from which it arose. And yet this unfamiliar source once struck man as strange and caused him to think and to wonder."

—MARTIN HEIDEGGER

In 2008, the movie *Iron Man* stormed onto the scene, raising a niche comic superhero into a cultural touchstone. The most memorable aspect of the series was not the action, which was pretty standard superhero fare, but the character of Tony Stark himself—genius, billionaire, playboy, philanthropist—and his famous laboratory. Much of the real technology we'll touch on in this book was represented in his lab: augmented reality, artificial intelligence, 3D printing, autonomous robotics, Internet of Things—all present and accounted for. What made the first movie especially fascinating wasn't the fantastic, but that it was just plausible enough to be engrossing. It was accessible science fiction.

Five years later, Elon Musk, a real-life billionaire backed by world-class experts and likely hundreds of thousands of dollars in equipment, shocked the world with an example of how one of his companies, Space X, quickly designed and fabricated rocket parts. He had a virtual reality headset prototype, hands-free gesture recognizing inputs, a customized 3D CAD system, and a 3D laser titanium printer. From direct virtual interaction with the digital twin of an engine, they could manipulate it virtually and print real metal parts on the fly to experiment with.

Six years later, I built my own similar workshop for a few hundred dollars, with a staff of zero. I bought an Oculus Quest from Best Buy, downloaded some CAD software, bought a 3D printer from Amazon, and built an IoT service connecting the two by following a YouTube video. I have no experience in CAD, no training in electronics, nor fabrication. I was able to use this setup to manipulate and print an emergency ventilator, following open source designs from Rice University in the early days of the COVID pandemic.

Let's think through this. In a bit more than a decade, we went from science fiction to science fact for techy billionaires to a consumer-available *Iron Man* laboratory, which required very little specialized knowledge and cost under a thousand dollars.

In this case and the rest we'll see in this book, there's a thin sliver of time—anywhere from five years to a decade—where an emerging technology is past the research and development phase, just on the cusp of scale, but not quite yet available for prime time. In marketing speak, you could call this the chasm between the early adopters and early majority. In Gartner's hype cycle, it's called the trough of disillusionment. In venture capital or technologist parlance, we call it *deep tech*. I'm fond of MIT Deep Tech Bootcamp co-founder Joshua Siegel's description:

A 'Deep' Technology was impossible yesterday, is barely feasible today, and may soon become so pervasive and impactful that it is difficult to remember life without. Deep Tech solutions are reimaginations of fundamental capabilities that are faithful to real and significant problems or opportunities, rather than to any one discipline.

Deep tech is both potentially disruptive and proceeds the more common "high tech" solutions. It is technology that, almost by definition, begs to be undervalued in its early days. But those who seize on the opportunity at the right time almost always end up the winners. If you grab on too early, you may find yourself as Yahoo or Friendster. Jump on too late, and you're Bing or App.net. But right on time? You're Google or Facebook.

WHY DEEP TECH MATTERS

We live in a world increasingly dominated by technology. Whether you're in finance, sales, design, logistics, or any number of fields and industries, technology is eating the world. Over a hundred years ago, factories were the cutting edge, powered by the assembly line, and the world was dominated by those who used them. Then came electricity. Then business structures like the firm. Then supply chain optimization. Then the world belonged to those who cleverly leveraged financial instruments.

Computers have been with us for decades, a tool in the background like hammers to be brought forth when we had a particular nail we needed to drive. But in the early part of the twenty-first century, something changed. Emerging technology ceased to create a competitive advantage. We now live in a fully digital age, and the major division today is between those companies who respond to that change and those who are left behind. Or as corporate tech expert Patrick Fisher said in Reuters, "all companies are technology companies now." The ante has been raised. Now the future belongs to those who don't merely adopt emerging technologies but invest in and drive their adoption, forcing everyone else to catch up—if they can. As we've explored, this earlier stage of technology is called *deep tech*, and in the decade between 2020 and 2030, seven technologies are poised to drive somewhere between 50 and 200 trillion dollars in new economic impact. It's not enough to survive anymore. Now the game is thrive or die.

IGNORING OPPORTUNITIES

Of course, there are those who don't believe tech will change the future, and some have suffered for it. Recent history is littered with corporations refusing to make the transition into leveraging technology appropriately, from the loss of a century of Sears dominance to the upstart of Amazon, to Hertz bankruptcy due to a billion ride-share cuts. In my experience, it seems almost fashionable in nontech circles to dump on technology. And I get it—honestly, that pushback isn't entirely unwarranted considering the endless stream of broken utopian promises rooted in technological triumphalism. But let's be sure we don't throw the baby out with the bathwater. The biggest danger in being ignorant of the current docket of deep tech's coming of age of is apathy, which in other words is a recipe for irrelevance. Whether you're beginning a startup, or you're a CEO or a thought leader, don't allow yourself to flirt with the lines of Luddite groupthink and be drawn into ignoring what you don't want to believe.

I was involved in several working groups in corporate and government environments during the early days of the COVID pandemic and was surprised at the steady march against contact-tracing technologies in favor of manual-only processes. Many of the arguments against adopting, for example, mobile apps to aid in contact memory or exposure notifications were pure Devil's advocate hogwash based on poor technology experiences from decades prior. One argument against technology that stands out in my mind was a belief that it was impossible for an algorithm to incorporate information beyond physical proximity, and only a human could provide other contexts and flexibility required in real-world use cases. I asked if the group was aware that AI has been making judgments at scale for nearly a decade and that humans weren't immune from bad judgment either. I was politely not invited to come back. The project was shelved for months, and an untold number of humans suffered for it. Only after a relentless push by scientists and technologists, it turned out that for every one hundred apps downloaded, one life was saved.

A major car company board member, quoted by McKinsey, once said: "The question is not how fast tech companies will become car companies, but how fast we will become a tech company." It's an interesting quote that sums up the greater and greater role technology plays in all business sectors, and how ubiquitous it's becoming. The boundary between being a "tech company" and "non-tech company" is blurring. As a simple example, in mid-2020, the top five market cap companies were all heavily invested in technology. The so-called GAFAM (Google/Alphabet, Apple, Facebook, Amazon, Microsoft) accounted for over $6.4 trillion and nearly 50 percent of the NASDAQ Composite.

You might think, "So what?" Tech companies are valuable, and I don't run a tech company. Here's a secret: Amazon isn't a tech company either. Sure, they have a website, but then again, you probably do too. Amazon is a marketing and logistics company that leverages technology really, really well. So well, in fact, that they built their own infrastructure and resold the sawdust to other companies called Amazon Web Services, or AWS. What about Google? Yes, they started as a tech company, but as Silicon Valley darling Peter Thiel has said, "Google is no longer a technology company." They're the world's largest advertiser. Facebook is, despite their apparent protests, a media company that deals in content streams down to the individual level. Apple and Microsoft? Apple runs their own television studios, and Microsoft is one of the world's largest video game consoles. That's not to say they aren't "tech companies," but they become less so by the day as they explore other lines of business. Tech isn't *what* you do; it's the underlying foundation to *how* you do what you do. And that makes understanding it integral to the business landscape.

THE WINKLEVOSS TWINS

If you want an example of the benefits of understanding deep tech, consider the Winklevoss twins. You may be familiar with these brothers, Cameron and Tyler, as popular foils in the Facebook creation myth. So the story goes, they were rich jocks with an uninspired idea for a social network and decided to sue the boy wunderkind Mark Zuckerberg for a piece of his sole creation. But despite how popular media had portrayed them in books and movies the early 2010s, the real story of these Harvard Gemini Olympians is one of redemption through a keen eye on deep tech.

In 2013, they saw the Bitcoin revolution coming. They hold degrees in economics but are not famous for being deeply technical men. Instead, they make it their business to be at the right place at the right time. They bought in on Bitcoin early, starting when the going rate was around $10 per BTC. Then they bought more and more, finally amassing a 1 percent stake of the total number of Bitcoin. They founded adjacent businesses, tried for an ETF (a mainstream instrument that would attract institutional investors), and created their own Bitcoin exchange. To support the technology and community, they worked as ambassadors for the power of decentralized digital currency. All the while, their investment grew. Once Bitcoin hit $10,000 per BTC in 2017, they both became the world's first Bitcoin billionaires. What a difference a decade makes.

If you're looking for new lines of business and revenue streams, process effectiveness, or other cost savings, deep tech is the most important avenue to investigate. Understanding deep tech is not merely a career booster but an important opportunity for raising capital. Funding sources for deep tech are exploding, and channels are multiplying. Venture capital and corporate funding are reliable sources eager to jump into the next big thing. Add in alternative funding like crowdfunding, initial coin offerings, and the newly found high-tech focus of sovereign wealth funds like the $100 billion Vision Fund, capital over the next decade is easily several trillion. Factor in the disruption to existing industries and the economic impact of the seven technologies that we'll dive into, capital will be available to capture a portion of the projected $50 trillion USD. This value is possibly an understatement, but I remained on the conservative side based on an analysis of several estimates from sources like Gartner, McKinsey, and World Economic Forum, or industry players like GE.

A QUICK NOTE ON ECONOMIC IMPACT

Economic impact is a broader concept than a simple addressable market. It includes primary, secondary, and tertiary affects. Selling IoT smart thermostats is a revenue generator for suppliers like Honeywell (primary), while using those devices can save their customers money on HVAC expenses (secondary), and finally, the aggregate reduction of energy usage globally can reduce carbon emissions from coal power plants (tertiary). So, in the simplest terms, I'm defining *economic impact* to be an addition to gross domestic product (GDP) across all countries, aka the gross world product (GWP).

So how can we land on figures a decade out when estimates vary wildly? By some accounts, AI alone will deliver $150T in new GWP by 2030. For comparison, in mid-2020, the aggregate of stock market capitalizations was $89.5T. Some figures place IoT over $100T on the high end and around $10T on the low. Following statistician Nate Silver's lead, I took a "poll of polls" approach, averaging estimates and weighting them according to the historical accuracy of the institution and the deep knowledge an institution has in a given space. I also adjusted large outliers against a clear converging consensus.

For example, GE's IoT estimates carry a bit more weight than smaller intelligence groups; meanwhile, *Business Insider*'s IoT estimates were a bit of an outlier from the consensus. Averaging collections of estimates tends to have a more accurate outcome than that of individual institutions. This is the so-called wisdom of the crowds. Without further ado, here is the macro estimate of all seven deep techs from 2021 to 2030.

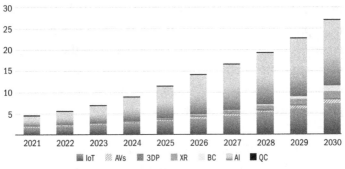

Deep Tech, Global Economic Impact (addition to GWP)

Legend: IoT, AVs, 3DP, XR, BC, AI, QC

A GWP estimate of deep tech over the decade of the 2020s.

"But wait!" you may say. "To sum the values in this chart, it's over $110T in impact over the decade." That's true, but the uncertainties surrounding any technology-adoption predictions over a decade cause huge error bars. So we'll be extraordinarily conservative here and take the worst-case estimates: a $50T opportunity. Best case, we're staring down the barrel of a future where deep tech spurs an additional $250T, or a quarter of a quadrillion dollars, to the gross world product. For comparison, according to Credit Suisse, all global wealth in 2020 totals to $360.6T. While I'm an optimist at heart, a lot of things will have to go right to bank on that figure.

WHAT TO EXPECT

Let's now dive into the nitty-gritty of what this book is about. You already know we'll be exploring seven deep technologies that will define the next decade and beyond. These seven technologies are general purpose technologies (*GPT*), which are those rare technologies that go beyond incremental improvements in a field or more than one field. Only a few dozen GPTs have emerged in history, such as the printing press, internal combustion engines, computers, and the internet.

GPTs are usually slow to emerge, although their presence has increased in rapidity over time. That being said, this decade is shaping up to be an extreme outlier—the sheer number of emerging GPTs is astonishing. We're looking at the emergence of *at least* seven general purpose technologies, each one capable of powering an economic revolution in terms of a significant percentage of gross world product.

1. Artificial Intelligence—AI, machine learning, deep neural nets, deep learning, and varieties including supervised, unsupervised, and reinforcement learning.
2. Extended Reality—Virtual reality, augmented reality, mixed reality, and things in between.
3. Blockchain—Cryptocurrencies for sure, but also smart contracts, DLTs, ICOs, DAOs, and uses for a trusted third party.
4. Internet of Things—Smart things, smart homes, smart cities, wearables, and the like.
5. Autonomous Vehicles—Self-driving cars and other AV-style robots.
6. 3D Printing—Mostly additive manufacturing but a bit of computer numerical control subtractive manufacturing as well.
7. Quantum Computing—Machines that will tackle the secrets of the universe that are impossible for today's computers to solve.

Each chapter will offer a deep synopsis of one of these emerging technologies. While you'll gain the most by reading all the way through, it's possible you aren't that interested in a given technology, or you're already an expert in it. That's fine. Each chapter is standalone and refers to other chapters only when discussing high-level convergence, such as the *intelligent edge*, which is AI living on IoT devices, or explaining that distributed ledger technology is just a rebranded name for blockchain.

In the chapters themselves, we'll cover a history of the technology (its emergence), examine possible use cases (which will be an exploration of how they're used now and how they can be used in the future), dive deep into how the technology works, and then end with current problems and challenges. Remember, this book is not a treatise on how to innovate. What I want to do is give you a foundation of knowledge that allows you to draw comparisons, identify trends, and dream up possibilities. I've kept the book as free as possible of roadmaps for the future or business ideas so that your mind is unhampered by suggestions and free to roam.

However, this book is not a stroll in the park. Some of these technologies are complex, like AI, while others are difficult because they're so broad, like IoT. Take the quantum computing chapter: it's a distillation of concepts learned from years of education in physics, specialized courses including MIT classes (thanks Prof. Oliver!), and hands-on experience with Microsoft, IBM, and D-Wave technologies. This is a difficult book, but you're smart and I believe in you. You'll get through it and suddenly find that you know more about this decade's bleeding edge of technologies than 99.99 percent of all other humans—a pretty exclusive club.

Ready? Let's dive in!

2

IT'S JUST ARTIFICIAL INTELLIGENCE

"AI is the runtime that's going to shape all of what we do."

—SATYA NADELLA, CEO OF MICROSOFT

The field of artificial intelligence is too nuanced for a definitive birth date, but one event in the late eighteenth century is a good starting point. In his attempt to impress the Empress Maria Theresa, a Hungarian inventor named Wolfgang von Kempelen built a chess-playing machine called "The Turk" in 1769. While other clockwork constructions were commonplace at the time, The Turk gained fame as a human-level intelligence machine that could win the game of chess. The secret of the mechanical Turk, sadly, lay in the fact it concealed a human inside the apparatus. It was a hoax, but for a brief time, it popularized the idea of a "thinking machine."

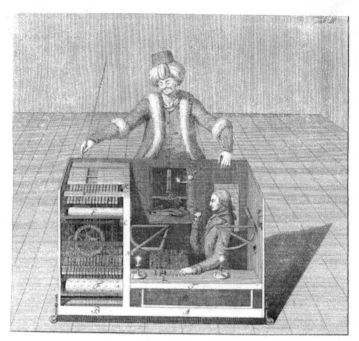

The Turk, image by Joseph Friedrich Racknitz, 1789.

Fast-forward to 1950 when Alan Turing, a World War II code breaker and a founder of modern computer science, penned a paper for the journal *Mind* titled "Computing Machinery and Intelligence." The seminal work introduced a concept he called *the imitation game*, referred to in later papers simply as the *Turing test*. He constructed the utilitarian notion that human-level intelligence is achieved when computers successfully convince us that they are thinking. As long as humans can't tell the difference, it doesn't matter if the machine really has human thoughts, whatever that means. What matters is that the machine can be useful in the same way as a thinking human.

For decades, computer scientists claimed we were on the brink of artificial intelligence (AI), but each of these claims ended in disappointment for those who yearned for the benefits of humanlike AI. In 1957, the first artificial electronic neuron was invented, dubbed the *perceptron*. While the press was eager to throw its weight behind this promise of AI, little came of this early research. Proofs were published showing fundamental weaknesses behind this form of perceptron, and progress crawled to an "AI winter" in the field, which continued until the late 1980s. Other forms of AI research continued, based on symbolic logic-breeding expert systems or genetic algorithms based on evolution or statistics based on the mathematics of fuzzy logic. Sometimes what passed for AI was a combination of techniques, but the achievements never fit the hype. There were just too many things that AI couldn't do, and humans could.

Then something changed. In 1997, IBM's "Deep Blue" beat grandmaster Garry Kasparov in chess. It was a fleeting affair, quickly dismissed as "not AI," since how generally useful is a machine that specializes in playing chess? The next few years saw machines beating humans at increasingly sophisticated games, but still, none could pass the Turing test. That is, until June 17, 2014.

Since Alan Turing first proposed his imitation game, computer scientists have battled in structured environments in an attempt to convince a panel of human judges they were talking to another human rather than a machine. That year, a program called Eugene Goostman convinced 33 percent of the judges at the Royal Society in London that it was a thirteen-year-old Ukrainian boy. In sixty-four years, it was the first time a machine had passed the test. And yet, it was also dismissed as "not AI."

Winning chess can be rejected as a clever mathematical trick with big computer hardware due to the relatively limited moves that need to be calculated to win a game. But there are games so complex that winning moves cannot be computed. One of these games is the ancient game of Go. A deceptively simple game, Go is played on a nineteen-by-nineteen-grid board with black and white stones. The complexity of calculating a single optimal move can take eons with modern computing infrastructure (sans AI), yet human brains can make expert moves in seconds. In March 2016, a program named AlphaGo bested the greatest living professional Go player in the world, Lee Sedol. Generalizing the technology allowed researchers to best humans at many other games with an improved machine called AlphaZero, and this wasn't another Deep Blue story. Critics were not so quick to dismiss this as "not AI." This time, something was different.

In May 2018, Google announced a project called Duplex. While conversational bots like Amazon's Alexa and Apple's Siri were getting better, they were still a far cry from natural conversation. At Google's yearly conference, Google I/O, the project leads played a couple of audio clips to an astonished crowd. The first was a humanlike computer voice calling a salon to book a hair appointment with a human on the other end of the line. The phone rang, a woman picked up, and the computer successfully booked the appointment. The second call was the same computer bot calling a restaurant to make reservations with another human. With a bit of struggle, the reservation was booked. What's more astounding is that in both instances, it was clear the humans at the other end of these calls had no idea they were speaking to a machine.

So currently, we have general purpose machines that can defeat humans in any number of games with fixed rules. We also have machines that can engage in open-ended conversations while synthesizing human speech such that a human at the other end of the line is unaware they're conversing with a machine. Machines are now doing many tasks that we have historically believed only humans can do, from *cognitive automation* in white-collar jobs to driving long-haul trucks and operating entire warehouses. It's time to realize we've achieved artificial intelligence or some form of it. Perhaps it's even time to recognize that we need a better definition of intelligence—one that no longer holds humans as a benchmark. There are certain tasks that computers do better than we can, and we're looking at a future where more and more of the "thinking work" historically done by humans can be taken over by machines.

HOW TO USE A NEW BRAIN

There are very few fields that artificial intelligence won't affect over the coming decades. Scratch that. Let's be bolder. Now that the Pandora's box of AI has been opened, we'll never stop finding new ways to add intelligence to dumb processes or inanimate objects. Look forward to a world of airplane parts that can detect mechanical stress and alert engineers of their need to be replaced, virtual doctors that diagnose patients from a science-fiction tricorder, and gimmicks like smart coffee mugs and AI assistants that automatically order your lunch based on your preference, budget, location, and history. Unlike other emerging technologies, cheap artificial brains can get smarter, coordinate with each other at the speed of light, and be cloned for free. AI is the most disruptive technology of the century, shifting the landscape of industries, old and new.

As an example of how a simple application of AI can transform a new industry, consider the case of smart parking garages. This relatively recent development helps drivers recognize how occupied a parking garage is by way of a green or red light above each space. You may have seen them at airport garages. Each of these lights is actually a self-contained computing device; it includes a sensor that detects the presence of a vehicle, and a small computer controls the system and shares its current state with a central network to provide an accurate count. This count is then displayed on the LED screens on each floor. If your smart garage has two thousand parking spaces, that's two thousand sensors that need to be purchased and maintained.

As an experiment, a UK research firm used their parking lot's existing security camera feed to train an AI to detect when a car was present in a space, then send the information through a cloud service to anyone with a mobile app. Since enough drivers already have smartphones, this system provides the same basic capability with no hardware investment. It's actually more flexible, too, since the model can get smarter over time, perhaps even predicting when spaces will open up based upon the habits of certain drivers. Building such a system is well within the capabilities of any computer science student in 2020. So, rather than investing in installing and maintaining thousands of physical sensors at the cost of hundreds of thousands to millions of dollars, software solutions can replace smart parking solutions for pennies on the dollar.

TASKS, NOT JOBS

Something to keep in mind as the AI revolution steamrolls human cognitive and manual work is that it's unlikely AI will automate entire industries in one shot. Instead, various tasks will be automated in conjunction with human counterparts. While an autonomous vehicle could certainly drive humans from a hotel to a conference center, those cars will still need to be maintained: filled with gas (or electrically charged), undergo oil and filter changes, and tires will need to be filled with air, patched, and rotated. Each of those tasks over time may fall prey to automation as well, but not all at once. In the interim, the number of tasks left to humans will drop or change.

As more tasks are automated, we're in for a strange disruptive period where those who have jobs will execute peculiar combinations of balkanized tasks. We're seeing it already. As systems for booking and managing travel become easier, the market for travel agents is shrinking. If you're a white-collar professional who travels, you're likely responsible for booking your own flights, cars, and hotels. You're possibly both a director of marketing and part-time personal travel agent. That's not necessarily a bad thing because travel satisfaction has increased steadily since people have been able to take matters into their own hands, not to mention the various industries created to support the democratization of travel. As Deloitte's *Future of Work* report states: "While some [people] will dramatize the negative impacts of AI, cognitive computing, and robotics, these powerful tools will also help create new jobs, boost productivity, and allow workers to focus on the human aspects of work."

It's hard to quantify exactly how many tasks exist in the world for AI to execute. But one way to get a handle on the changes happening now is to focus where AI automates workloads by sectors that constitute huge portions of the economy and workforce.

AGRICULTURE

Historically the largest industry, agriculture has long been in the crosshairs of innovative technologies. From plowshares to cotton gins to factory farming and GMOs, each innovation seems to increase yield and decrease the number of people required to work in this grueling sector. In 1870, more than 50 percent of Americans were employed in agriculture. Today, that figure is under 2 percent. AI is expected to continue this trend of fewer employees to greater yield.

The perennial pestilences of farming have always been a combination of invasive plants (weeds), unwelcome wildlife (from insects to rabbits), unpredictable weather (droughts and floods), and large-scale monitoring (soil and crop). We are in the midst of a gold rush of AI-based solutions to these issues.

Over 250 insect species have become resistant to chemical herbicides, and excess chemical runoff into our water supply has an environmental impact. John Deere announced the acquisition of a company that leverages machine-learning vision systems to automatically spray weed killer directly onto plants, reducing herbicide use by 90 percent. Several other companies, such as ecoRobotix, are creating chemical-free, mechanical weed-pulling robots. Many of these robots are also capable of targeted insecticide deployment, helping stave off many of the unintended consequences of overspraying, such as bee colony collapse. And speaking of bees, there's now a pollinating robot called BrambleBee.

Nearly 90 percent of crop losses are due to weather-related events, and the task of weather prediction is tailor-made for Big data and machine learning. The United States National Oceanic and Atmospheric Administration (NOAA) has increasingly accurate hail prediction, allowing farmers to better map out planting and harvesting. HydroBio is a company that leverages hyperlocal data and AI prediction to help farmers know when they should irrigate. A handful of other longer-term climate systems can help agribusiness decide which parts of the world will be safe to plant in over the coming decades.

A solar powered weed robot.

Monitoring all the details of million-hectare farmlands is daunting work for humans yet tailor-made for machines. With deep learning systems like Plantix, which uses satellites to track soil erosion and nutrient levels, computers can tirelessly detect soil issues at scale. For a more real-time, fine-grained look, a company named SkySquirrel pairs machine learning with an army of drones to detect invasive plants and even molds at scale.

MANUFACTURING

Walter Reuther, powerful leader of the automobile workers union, and Henry Ford II, head of the Ford Motor Company, once toured a new factory filled with a line of automated robots that were building cars. After a brief silence, Ford asked, "Walter, how are you going to get those robots to pay your union dues?" To which Reuther replied, "Henry, how are you going to get them to buy your cars?"

This apocryphal exchange occurred in the early 1950s. Even then, automation was taking over manufacturing tasks, and this story showcases a persistent fear of automation that's been with us since the Luddites of 1811 first destroyed high-tech cotton mills. Relatively few jobs in first world nations today are in manufacturing, as economies give way to the service sector. Still, over 8 percent of Americans earn livings in the manufacturing industry, which is over 11 percent of US GDP. Much of the technology needed to automate some of these jobs currently exists: robotic arms, logistics machinery, quality control systems, and the like. AI turns expensive specialty robots into general purpose cobots. Rather than huge, clunky welding robots, blind to the world and programmed for a narrow range of tasks, cobots can be taught many different tasks, retooling themselves automatically. Cobots are also aware of their surroundings, capable of working side by side with humans on complex tasks. This allows cobots to slowly ease their way into a workspace and take over more work, limited only by an exponentially growing intelligence. The most famous cobot on the market is Baxster and costs around $50,000 US. It may not be able to do all tasks, but it can do enough to bend that 8 percent of jobs down a few points.

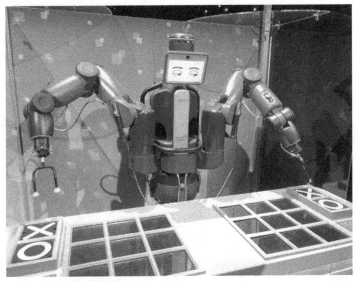

A Baxster cobot playing the author in tic-tac-toe.

Outside of the factory, AI has made vast inroads in the realm of logistics. The ability to finely track and trace shipments allows for less waste and more mobility of products, reducing inventory on hand. Many warehouses are nearing peak automation, be it Amazon in the US or JD in China. The next step is autonomous shipping, which Waymo and Uber are working on. Reduced waste, reduced cost and overhead, and faster time to market is good for companies and for consumers.

MILITARY

For the military, AI poses a siren song that's too attractive to ignore: perfect knowledge of world events from governments to battlefields, paired with robots that bend the casualties on your side toward zero. AI can better support troops by improving training systems and creating novel curricula for war games. It can provide smart weapons and better intelligence, along with the more pedestrian benefits of industry, like optimizing logistic challenges in the world's most challenging situations, or helping troops with maintenance tasks. Natural Language Processing (NLP) can take on many roles of human translators today, and NLP at scale can sift through vast amounts of audio surveillance in real time. Furthermore, complex associations between unrelated datasets will stitch together a single narrative. Imagine tracking the movement of a suspected terrorist using various video feeds that leverage facial recognition, cell calls with voice recognition, and software that cross-checks travel documents through known aliases.

Autonomous weapons are increasingly augmented with AI, such as smart-camera-controlled tactical missiles. Even if control ultimately remains in human hands, the myriad of complexities that would take humans years to learn can be partially automated, allowing operation from fewer specialist hands, like flying attack drones. Moreover, the ability to correctly detect targets can drastically reduce collateral damage and innocent deaths.

These are only a few straightforward examples and may not even scratch the surface of the many uses for AI in the military. At the very least, increasing automation may allow countries to shrink their military budgets in favor of more civilian expenditures.

SERVICES

Agribusiness, industry, and military are important for a functioning society, but most first-world economies operate primarily in the services sector. Services accounted for 82 percent of US GDP in 2018 and employs most working adults. This is also the segment that is most impacted by AI. Why? Because, contrary to common belief, cognitive labor is the easiest sort of work for AI to execute. While manual robots still resemble disturbingly precise drunks, it's relatively easy to automate the process of updating spreadsheets. The growing field of *robot process automation*—the discipline for automating manual workflows—has only marginally lived up to the hype, appearing only in corner cases of office environments. However, its AI-powered cousin, cognitive automation, is poised to disrupt many tasks that require sitting in front of a computer on a daily basis. These range from white-collar office jobs to medical to finance and even to creative work, like writing and music.

OFFICE JOBS

Office jobs are great. You get to be indoors, sit in a comfy chair, and the schedule tends to be pretty consistent and reliable. Perhaps I'm giving away too much about my own personality, but the weather is nice and physical exertion tends to be minimal. And despite the wide range of specialties, office work has similarities no matter your title or expertise. You'll likely work with a computer or other device like a phone; compose and answer emails, texts, or instant messages; create, fill out, or adhere to a schedule; and shuffle information from one location to another, even if it's in the form of just answering coworkers' questions.

AI makes most of this easier. One of the first AI assistants I tried was X.ai, and I was immediately hooked. All of the coordination often done by administrative assistants was automated with this email-based chatbot that understood nuanced human speech and could converse clearly and politely with humans. This same bot had knowledge of your calendar, travel times, and blackout times (like lunch from noon to 1:00 p.m.). If you're a manager, there's an AI called Workloud, which tracks the actual attendance of meetings in terms of scheduling; AI Sense, which transcribes verbal conversations at meetings into searchable text; and Allocate, which tracks timesheets based on any real work done, saving your team the task of self-reporting.

Hopefully, your day job is more than just attending meetings. As you do actual work that interfaces with a computer—from taking and entering orders, fixing issues based on consumer requests, or other data entry work—robotic process automation (e.g., Pega) and now cognitive automation (e.g., Corseer) can be trained to execute a wide variety of tasks automatically. Of course, this assumes you actually know what your process is. There's an AI for that. Celonis can plug into your digital communications and processes, map what you think your process is, what it actually is, and how close your reality meets expectations. Moreover, lots of office work consists of answering questions whose details are either in some system or in an employee's head. The ability to catalog and make this information available for those who ask in a natural way, from emails to phone calls, is well within the capability of several AI knowledge systems called chatbots, from Spoke to GrowthBot.

AI is great and all, but for the foreseeable future, we'll still need people. Surely only people can find other good people, right? Well, AI is coming for human resources too, pal. From writing job descriptions (Textio) to talent searches (Koru) to screening résumés (Ideal) to onboarding and training (Chorus), more and more tasks that fall to HR are being done cheaper, faster, and better with a cadre of well-placed machines. On the job-search side, there is a growing cadre of AI tools designed to help you craft a résumé that can help navigate the gauntlet of bots on the demand side (Jobscan.co).

A German online retailer, Otto, currently uses AI to predict whether or not a customer is likely to purchase a product with 90 percent accuracy and start the shipping process before the order is placed. This not only reduces wait times, leading to happier customers, but also saves the company millions of dollars a year in excess inventory.

We could go on and on, and tear through every office function. There are myriad solutions for sales (Salesforce Einstein), marketing (Albert.ai), public relations (Signal AI), legal (iManage), finance (Squirro), technology operations (Moogsoft), risk (Exabeam), security (DeepArmor), and so on. What's important to keep in mind is that each of these specialized solutions will continue to improve, generalize, and take on increasingly sophisticated tasks. This is because computers continue to exponentially improve, unlike our slow, squishy human brains.

MEDICINE, LAW, AND FINANCE

For generations, parents hoped their kids would end up as doctors, lawyers, or bankers. These were secure, well-paying jobs, requiring intelligence and education. But it turns out that even these professional careers are not immune to the slow onslaught of the AI revolution.

Worldwide, around 10 to 15 percent of a country's GDP is spent on healthcare. In the US, thirty cents of every dollar spent on healthcare goes to waste and administrative overhead. If every healthcare office adopted even standard office automation, some percentage of that overhead would decrease, thus materially shrinking overall healthcare costs. But many wastes are specific to the medical field, from excess lab services to missed prevention opportunities. While it's unlikely that doctors or nurses will go away anytime soon, many of the specialized tasks that require the most schooling are changing thanks to AI.

While IBM's Watson began life as the greatest *Jeopardy* world champion in history, it has changed careers and is now a world-class medical diagnostician. Recently, China's best brain cancer specialists lost in a diagnosis competition against an AI. On the side of paperwork, insurance risk assessments and the tasks of billing and coding (H2O.ai) are being automated, as our over-worked doctors and nurses will need this help as Boomers age. On a quality note, R&D is improving as AI can stay current and ahead of an endless avalanche of emerging publications. All of this automation leads to fewer errors, lower costs, and freeing up time for more personalized medicine.

The same changes are happening in the legal sector. Despite what TV shows would have us believe, very little of the work lawyers do takes place in a court room. Whether it's a district attorney building a case with police evidence or a corporate law firm digging through client records, much of legal work consists of a process called discovery. Discovery is perfectly suited for computers in general, and for AI in particular (for example, tools like Exterro). Computers never get tired or distracted, and can make meticulous connections in large swaths of data far beyond the abilities of humans. "If I was the parent of a law student, I would be concerned a bit," says Todd Solomon, a partner at the Chicago law firm McDermott Will & Emery. "There are fewer opportunities for young lawyers to get trained, and that's the case outside of AI already. But if you add AI onto that, there are ways that is an advancement, and there are ways it is hurting us as well."

Finally, banking: AI has been infiltrating the world of finance for years. High-frequency trading (HFT) and risk assessments have been run by algorithms for decades, and those algorithms are being increasingly sharpened by AI. Cognitive automation has been handling the work of acquisitions, which was, until recently, a relatively high-paying and well-respected job due to its complexity. And bank processes are becoming easier—for example, both Chase and Bank of America allow you to deposit a check simply by taking a picture of it with your smartphone and having it verified by AI.

Like any sector of the economy, professional jobs are being reshaped by AI. It turns out the jobs that require the most training and knowledge are just as susceptible to automation as office or manual work. And maybe more so, considering the incentives to get the work correct, and the great cost of employing humans in these fields.

THE ARTS

Many believed that creativity would always be outside the realm of AI. But that myth is slowly being debunked as AI infiltrates the realms we've always considered to be the arts: writing, drawing, and music. In journalism, for instance, AI is making its presence known. According to a *New York Times* article written by Jaclyn Peiser, over 30 percent of *Bloomberg* articles were being written by AI at the beginning of 2019. And Kristian Hammond of *Narrative Science* estimates that 90 percent of major newspaper articles will be written or assisted by AI by the year 2030. This started with a program called Quill, a sports article generator, which now reports on finance and global events for most major media outlets, unbeknownst to most of its readers. While there will always be a place for *Atlantic*-style think pieces, most of the news we consume is far more mundane and merely conveys information, like recapping a State of the Union address, with an occasional quip that requires little human intervention.

But we've already seen that AI is really good at dealing with text. What about the other arts? Cambridge Consultants created a *generative adversarial network* (GAN) called Vincent that allows anyone to sketch a picture, and it will generate a painting based on what the user is trying to convey. There are apps that even convert speech ("I love tabby cats") into visual art (a cat decorated with hearts). This doesn't replace human creativity but is augmenting it and placing artistic expression within the reach of every human, regardless of technical skill. And while composing music with AI has been around since Ray Kurzweil tried it in 1965, AI-composed music has more recently found its way into the mainstream with the release of singer Taryn Southern's computer-generated album called I Am AI. Generating art and music has become such a common pastime for researchers that Google created an open source AI art platform called Magenta, allowing nonspecialists to experiment with generative art.

AI is taking over agriculture, industry, military, service, professional, and creative tasks. You might be forgiven for taking the stance of Elon Musk or Stephen Hawking and believing we're scant years away from an automation-fueled uprising. But before we freak out and welcome our new robot overlords, let's take a breath and see how this AI thing actually works. Maybe then we can revisit the glaring weaknesses in the system and find some solace in the fact there are still plenty of jobs we humans are qualified to do for the foreseeable future.

WHAT'S DIFFERENT NOW?

Fueled by a boom in a branch of AI called machine learning (ML), AI has emerged from being an academic curiosity to changing the world in practice. How did we get here? A handful of shifts in the landscape of technology has contributed to this AI resurgence over the past decade, and these core investments are poised to keep bringing new cognitive capabilities to bear on an ever-wider array of goods and services. These improvements have risen from three coequal changes: better hardware, democratization of algorithms and data, and increased investment in the ML space by both industry and academia. The changes in the industry are symbiotic. For example, on March 2017, Google's CEO announced that they were now an "AI first" company, but they had invested in ML research for over a decade. Internal investment, academic partnerships, and acquisitions set the stage for Google's modern AI renaissance. Concurrently, they invested in an open ML toolkit called TensorFlow in an attempt to attract top talent and control the narrative on new applications. This coincided with the creation of custom hardware specialized to execute tensor calculations called TPUs (Tensor Processing Units). Having the world's largest collection of data didn't hurt either.

BETTER HARDWARE

While there are many ways to implement artificial intelligence, the current growth is based on machine learning in large part because machine hardware has improved in comparison to other artificial means, such as synthetic biological research. While Moore's Law might be dead (stating that transistor density doubles every eighteen months—more on this in the "Picking Quantum Locks" chapter), the kind of specialized hardware necessary for executing modern machine learning techniques continues to grow exponentially, thanks in large part to two unrelated trends: video game enthusiasts and cloud architectures.

The kind of hardware necessary for executing machine learning systems is similar to the hardware optimized for rendering video game graphics, called graphics processing units (GPUs). This is similar to the central processing unit (CPU) that has dominated the computing market for decades but is optimized for the kind of math operations necessary for both use cases. With the proliferation of cheap and easily available GPUs, new life was breathed into the stagnant field of ML research. Google's play in this space upped the ante, as they rapidly prototyped, built, and deployed their first generation of production TPUs by 2015. The race was on to create specialized machine learning hardware called *AI accelerators*. Nvidia, the world's leading GPU manufacturer, staged a concerted effort to dive into the greater AI market outside of Google's ecosystem.

While the AI hardware war rages, the cloud is the primary battleground—namely, Google Cloud, Amazon Web Services, and Microsoft Azure. The cloud allows anyone to access cutting-edge hardware.

DEMOCRATIZATION

While it's true that Google was an early mover in the ML space, most of the concepts and tools they championed were developed elsewhere and, in many cases, were better than theirs. What Google should be appreciated for is forcing other technology companies to open their IP. For example, Facebook, IBM, and others contributed to the core project of an open source ML library called *Torch*, which was released in 2002, but were reticent to release how their particular "sausages were made." But Google released more—not only the TensorFlow in 2015, but also a mountain of documentation, training videos, blog posts, and open-sourced algorithms that worked in production environments. The rapid popularity of TensorFlow ushered in an age of open, sharable machine learning based on a common language.

This openness in software, coupled with newly accessible, powerful hardware, democratized the machine learning ecosystem. Suddenly, any smart kid with a cool idea had access to cutting-edge ML research, with source code and an environment to run it in. Another driver behind the democratization of machine learning is the flood of easy-to-use frameworks like the point-and-click Orange or SageMaker, coupled with easy-to-grasp education opportunities like Udemy or edX. No longer was machine learning reserved for those with a "freakish knack for manipulating abstract symbols" (via Paige Bailey of Microsoft, quoting Bret Victor[1] at an AI conference). But open AI algorithms are only half of the story. Machine learning requires data to train with and run against, and data is increasingly available everywhere.

The proliferation of open data, from public weather to university psychology research to government census and economic data (data.gov) has given many professional and budding AI engineers a set of data against which to build their own ML models. Often, those models are open sourced themselves, allowing others to build on their work, and prompting organizations to open even more datasets. This virtuous cycle of data to information has created new understandings in previously opaque

---

1   Bret Victor, "Kill Math," blog published April 11, 2011, http://worrydream.com/KillMath/.

industries, which prompts even more open data, democratizing machine learning even more. There is a constant stream of competition to create the best AI against a given dataset hosted on a site called Kaggle. Kaggle provides the data and terms of the competition, and a dispersed community of data scientists compete to make the most sense of the data. These competitions often have a cash prize for the winners and range anywhere from "Customer Revenue Prediction" to "Using News to Predict Stock Movement" to "Human Protein Image Atlas Classification," and those were all in the same week!

Hardware as a service, machine learning lingua franca, open-sourced algorithms, open datasets, and easy-to-access education all play a role in the democratization of machine learning. It's great that this infrastructure is opening up, but why would companies like Google give everything away? Where does the money come from for the Kaggle prizes? Who pays for all of this?

INVESTMENT

Those of us who were around for the big data revolution quickly spotted a flaw in our operating model. While the emerging big data industry made it easier to quickly collect huge volumes of varied data (for example, by leveraging NoSQL databases, which helped make this era possible) it was not easy to make sense of the data. The general philosophy for the better part of a decade, starting around 2009, was "Collect all the data; we'll figure out what to do with it later." Turning data into information is a difficult task. Turning information into understanding at scale is nigh impossible for humans. A new, vaguely defined kind of job started popping up everywhere: data scientist. It was no longer good enough to hire statisticians; these unicorns also needed to be experts in large-scale data management and mine mountains of unstructured data for...something. The company made an investment in all of this data just to find something good.

Everyone had a sense that data was valuable, but there was no clear roadmap on what to do with that data. It was like a joke from the animated TV show *South Park*, where gnomes secretly stole underpants with no clear plan in sight: step one, collect underpants, then step two, then step three is profit. There never was a step two, but they felt very strongly that it would lead to profit. We liked to believe the coterie of swells slinging real money had clear goals in mind, and sometimes it was just worth taking a shot. It was highly unlikely the data would be worthless. The emergence of ML as an increasingly popular subdiscipline of data science ended up being the "step two" to make sense of all the data sitting largely idle around corporate data centers. The rapid increase in industry and government spending on limited AI resources makes sense through this lens, alongside the standard concerns about being left behind.

The symbiotic nature of academic and industry research is important for emerging technologies like machine learning. Academics conduct leading-edge research across a variety of topics, and some percentage of that research ends up, hopefully, being of interest to the corporations and governments of the world. They, in turn, invest in more of the kind of research they believe will yield better results. Right now, there's a gold mine of investment, and it's only growing. Corporate investment in AI is on pace to have around 50 percent CAGR (compound annual growth rate) over the next decade.

TYPES OF MACHINE LEARNING

AI continues to evolve from atavistic statistics that merely describe the world to making sophisticated predictions to prescribing courses of action for humans to eventually taking action itself. The tools required to evolve these capabilities have become increasingly humanlike in the manner in which they understand the world. While the perceptron was insufficient for much of anything with the technology of the 1950s, artificial neurons are the design of many modern AI via ML and take the form of deep neural networks. In a short time, AI has moved from simple data structures and symbolic algorithms to a complex artificial neurology, built on structures of available data and hardware.

PLAIN OLD DATA SCIENCE

In the most basic terms, statistics is about creating models that estimate unknown parameters. Inference, probability, frequency, data science, machine learning, artificial intelligence, human intelligence...they are a rogue's gallery of approaches to attacking a basic problem: perfect knowledge is not possible. So, how do we fill in the gaps when confronted with something new?

Statistics is as old as the first humans making assumptions based on observation. The Greek historian Thucydides described a frequentist method used in the fifth century BCE. One of the first modern statistical models tracked mortality by sampling signaling for the bubonic plague (John Graunt). Around the same time, the study of randomness was being used to calculate probabilities in games of chance (Blaise Pascal). Statistical work for demography and probability theory started to converge (Pierre-Simon Laplace) over the centuries, eventually consuming warring philosophies like frequentists and Bayesians into a general set of methods.

With the emergence of computers and the big data revolution, a new field of data science arose. It basically consists of statisticians who know how to handle lots of data and can code computers a bit. Like other scientists, they test hypotheses against large datasets, using their own tools of the trade—namely, software packages like R or Python's scikit.

A good example of a useful statistic for extrapolating signals from incomplete knowledge is one of the simplest and oldest, called linear regression. Imagine you had a bunch of measures of people's height compared to shoe size. While each of the dots in a 2D chart represent a single person, a pattern emerges over a population of people. Generally, the taller a person is, the larger his or her feet will be. Now, let's draw a line through what appears to be the average of each value. This line is our *prediction*. Say the average 60-inch tall person wears a shoe size of 8, while a 70-inch person wears size 11.

But is our prediction line accurate? To find out, we can measure the distance from each point to the line. That distance is called the *error* because our estimate is incorrect compared to the real observed value. We then square the errors (to make distant dots stick out even more) and add them all up. That total is called the *sum of squared errors* (SSE). We want to draw a prediction line that best fits the data points we have with the lowest SSE since that means our line is closest to the most points overall. Linear Regression is a method for calculating the best prediction line to the dataset.

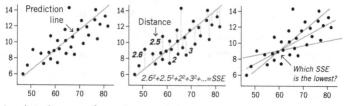

A prediction line, a sum of squared errors, and a linear regression.

Although we'll never have a line that exactly predicts every new measurement, we only need one that is close enough to be useful. That's the crux of data science right there. We aren't looking for perfect; we're looking for useful. There's also no law that says you have to group data points together using straight lines. We can also try curved lines, aka nonlinear regression, as would be the case if we measure average height by age. After people reach age eighteen, the correlation between height and age tends to flatten out, and even curves down a bit later in life.[2]

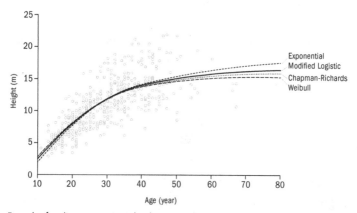

Example of nonlinear regression in height compared to age.

Sometimes we want to analyze clusters of data (K-means) or reduce the dimensions or features in play (manifold learning)

2   Park, et. al., "Height Growth Models for Pinus Thunbergii in Jeju Island," *Journal of Forest and Environmental Science* 31, no. 4: 255-60, DOI: 10.7747/JFES.2015.31.4.255.

or convert from one type of data to another (autoencoding). Moreover, not all data is numeric. Sometimes we want to classify things, such as "Is that a picture of a cat or a dog?" The tools for accomplishing this feat are varied, including logistic regression, K-nearest neighbors, support vector machine (SVM), decision trees, random forests, and so on. How can a data scientist possibly know which algorithms to use, let alone figure out how to fit the chosen model to the dataset?

The answer is we start letting the machines do the work for us. We educate a model with a training dataset and look to reduce its errors against a validation dataset. Then we check how generally useful that model is with a *test* dataset. In other words, we teach the machine how to learn by fitting a prediction curve with the common pattern of training, validation, and testing.

DEEP NEURAL NETS ∩ MACHINE LEARNING = DEEP LEARNING

"Excellence is an art won by training and habituation...we are what we repeatedly do."

—ATTRIBUTED TO ARISTOTLE BY WILL DURANT

Artificial neural networks (*ANNs*) are loosely inspired by biological neocortex neural clusters. Deep Neural Networks (*DNNs*) are ANNs where there are many layers of neurons between the input and output—in other words, the network is deep. *Deep Learning* is basically training a DNN with loads of data, until the network starts to be "shaped" by the commonalities in that dataset. Let's say you train a DNN with many images of cats. With enough images, the DNN will start to recognize the common attributes that make up a cat. When you give it a picture it hasn't seen before, it can pick out whether or not the image contains a cat with some level of confidence. Machine learning is not magic; it's just multidimensional curve fitting.

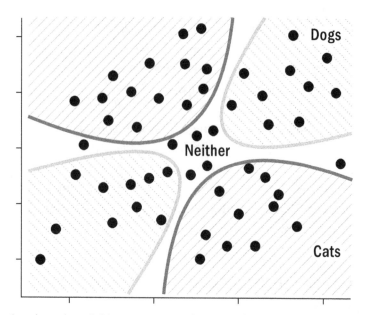

Some dots are best called dogs, some are cats, and some are neither.

Consider an audio waveform. Since sound is just vibrations in the air, it was discovered long ago that unique sounds can be collapsed into one unique wave. Let's say we have a series of waveforms of different people saying the word "hello." Something about the waves are similar enough that any human capable of hearing could make out the word "hello" (even if they don't speak English). While the waves might not look exactly the same, there is some commonality that can be extracted by a crafty algorithm.

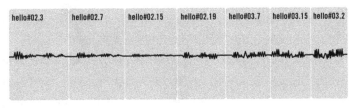

| hello#02.3 | hello#02.7 | hello#02.15 | hello#02.19 | hello#03.7 | hello#03.15 | hello#03.2 |

You train a DNN to recognize "hello" by giving it many samples.

Send many samples of the audio waveform for "hello" to a DNN, and it may, through a series of weights and adjustments to its internal structure, start to recognize the pattern in any waveform. Then, as you stream in a series of waveforms of, say, a conversation, it can pick out a particular wave segment as the word "hello." This vaguely mimics the way our human neurons recognize a friend shouting "hello" across a crowded room.

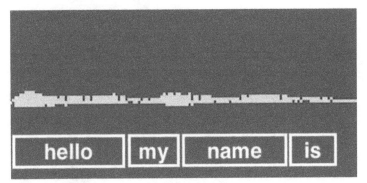

The trained DNN model can recognize words from the waveform.

The ability to recognize patterns in a set of noisy data is similar whether it's through audio, images, video, electrical pulses, financial data, or many other signals. The mathematical representation of these data is called a *tensor* (think of it as a matrix containing other matrices). As long as the data can be converted into a tensor, it's a candidate for deep learning, and it turns out, almost everything humans interact with is also.

DNNs and Deep Learning open up a world of ML techniques for various uses. The "hello" example was a way to classify an observation and is becoming a rather milquetoast technique in the ML space. Given the right incentives, you can teach machines to do more than just observe; you can also teach them to take action.

SUPERVISED, UNSUPERVISED, AND REINFORCEMENT

The examples we've covered generally fall into a category called *supervised machine learning* (SL). The word "supervised" here means our training data has a value (dependent variable) that represents what we're training the AI to predict. If we want to train a convolutional neural network (CNN) to recognize images that have cats in them, we have to train it against many other images of cats that humans have already labeled as "cats." If you've ever run across a modern Google CAPTCHA (security) test, you may have been provided a series of images and been told something like "Select images containing a stop sign." That's Google using you to label its images so it can later train a DNN to recognize stop signs automatically. You, dear human, are the machine's supervisor.

This raises the question: Is there an *unsupervised machine learning* (UL)? When you have a lot of data and want to detect patterns but aren't sure what those patterns might be yet, you can't label it. Imagine you have a series of warehouses, where inbound products are scanned and their geolocation is tracked. Providing only these locations, certain algorithms (isolation forest, K-means clustering, variable autoencoders, etc.) learn where your products are expected to be. If they come across a scan of a location that's unexpected, like Antarctica, this can trigger an alert that a signal is an anomaly. Unsupervised machine learning is commonly used in many kinds of anomaly detection, from logistics to bank fraud to consumer profiling and recommendation systems—anytime you're looking for a good representative form from the data.

Recently, ML practitioners have been playing with combining SL components with UL into a type of ML called *generative adversarial networks* (GAN). We could call GANs semisupervised models because they're trained by arming an unsupervised model to learn (and generate) against a known supervised model (the adversary). We won't go deep into details here, but GANs are interestingly powerful: they take bodies of known works and generate outputs that are similar enough to be useful. For example, we can provide a corpus of classical music and ask the GAN to generate an endless supply of new music that somewhat "sounds like" the input. Well-trained GANs are excellent for generating new creative endeavors, such as an endless selection of custom sneakers. In early 2019, an OpenAI GAN was claimed to be so good at generating convincing fake news articles, it was deemed too dangerous to release to the public. To much fanfare in the AI community, it was later released in 2020 under the name GPT-3, and immediately used to generate convincing fake blog posts that went viral.

In the history of psychological behaviorism, Pavlov's dog tends to be the experiment most of us would shout out at a trivia night. But years before Ivan Pavlov rang his feeding bells, Edward Thorndike discovered the "law of effect," which states that satisfying consequences tend to be repeated, giving rise to a study called *operant conditioning*. This is the crux of a third type of machine learning, called *reinforcement learning* (RL). Unlike supervised or unsupervised machine learning, reinforcement learning goes beyond data per se and instead focuses on training *agents* to act in a given *environment*. We train these agents the same way we tend to train other humans: by rewarding desired behaviors and punishing undesirable ones in service of some goal. Reinforcement learning, as a computer concept, is decades old. However, RL is experiencing a renaissance thanks to the emergence of deep neural networks, most famously Deep Q-Networks (DQN). RL tends to be the modern tool of choice for training machines to do things that have been done by people, such as beating world champions at Go (AlphaGo), or training computers to trade stock autonomously better and faster than any human could with the goal of higher profits. Of all the tools in the roboticist's toolkit, reinforcement learning may become the most disruptive.

When you tie together RL, SL, and UL, you can get a sense of the next few decades of AI research. Consider autonomous vehicles. You could use cameras and CNN (SL) models to see and detect objects, use an isolation forest (UL) to judge whether the objects together make common sense ("Is that a snowman on a palm tree?") and DQN (RL) for a car to react based on the best available knowledge to drive down the road without hitting a living thing or being hit. While cutting-edge researchers continue to build increasingly sophisticated ML models, we mortals can put them together like puzzle pieces in new and interesting ways.

PROBLEMS, OR THE KOBAYASHI MARU

Despite the promise of artificial intelligence powered by machine learning and the rapid gains over the past decade, there are a few issues that the industry must hammer out before we can safely coexist with our own personal Rosie the Robot housekeepers. AI is not only complex to implement, but unlike code, it's not easy to check under the hood and see what's going on. This makes many industries with tight quality controls nervous. AI rejection is exacerbated by the same institutional inertia that slows any new idea: "We've been doing fine for years." Combine known biases in data, economic and philosophical concerns, and the existential threats of general artificial intelligence, and we have a lot of work to do before unleashing this particular cat from the bag.

COMPLEXITY

"We are now writing algorithms we cannot read. That makes this a unique moment in history, in that we are subject to ideas and actions and efforts by a set of physics that have human origins without human comprehension."

—KEVIN SLAVIN, RESEARCH AFFILIATE AT MIT MEDIA LAB

Deep neural nets are notoriously difficult to unravel. It's not easy to ask the computer how it landed on a particular decision. Beyond the complexity of developing ML models, using them can be equally frustrating. There are many cases of deep image classifiers that can be fooled with a little bit of noise in a photograph. These are imperceptible to humans, but can fool an AI to believe, for example, that a picture of a bus is actually an ostrich.[3]

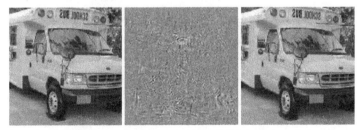

An image of a bus superimposed with noise can make an AI think the last image is an ostrich.

---

3   Christian Szegedy, et. al., "Intriguing Properties of Neural Networks," *International Conference on Learning Representations* (2014), https://research.google/pubs/pub42503/.

But as research into ML continues, these attacks are being accounted for and baked into increasingly robust models. It turns out the kind of attack that can generate noise can also be used to mitigate it, sort of like inoculating the AI against known viruses. In the future, we'll have to design a sort of "immune system" to recognize and inoculate against tricks of increasing sophistication.

A subtle but important issue is the emerging "reproducibility crisis." While AI techniques help scientists break new ground in complex fields such as protein folding (a DeepMind AI took home the gold in an international competition in 2019 between the world's top scientists), many victories are pyrrhic. A cornerstone of science is the ability to reproduce results. If we can't retrace the process by which AI reaches its conclusions, it puts any AI-based theories on philosophically shaky ground. This has launched a new field of AI research known as explainable AI, or XAI. Whether XAI will yield practical results is yet to be seen, but initial research is promising. But until we're able to peer into the brain of AI and understand how it comes to its conclusions, we have to tread carefully.

BIAS

In January of 2020, a private citizen named Robert Williams was arrested on his front lawn in a quiet Detroit suburb after a facial recognition AI wrongly labeled him as a match for another man caught on camera in a robbery. He was innocent. The police trusted the system so completely they took this algorithm as strong evidence and believed they found their man. After all, how can a computer be biased? It turns out, several studies have shown AI can be quite biased. Cases can be as severe as false identity to as subtle as a top social network's image cropping algorithm disproportionately favoring certain people. In fact, AI models adopt the very systemic biases of racism and sexism that plague human societies.

In early 2018, a study of the top gender-recognition AIs from Microsoft, IBM, and Megvii correctly identified a person's gender 99 percent of the time—as long as they were a white male. Conversely, dark-skinned women were only accurately identified 35 percent of the time. The discrepancy is due to the images chosen to train the ML models. While there's little reason to believe racial intent on the part of the data scientists, the unconscious bias of the dataset they chose came through in the model. Using these biased models for profiling humans in a production setting potentially leads to false identifications, arrests, and punishments, all based on the testimony of a machine that we believe could not possibly hold prejudices. But in reality, if we train an AI with a dataset of people who have historically been wrongly accused as criminals, the AI will continue the pattern. Train an AI against traits like credit worthiness, and redlining will continue for generations—we'll just be calling it by a different name.

The good news is that there is now a concerted effort between academia and industry to discover and account for biases in ML models and training data. The *MIT-IBM Watson AI Lab* is one of the groups working on improving the technical side, while advocacy groups like *AI Now* in New York are warning officials of the dangers of taking AI outputs at face value, especially when the claims strongly disagree with our human intuitions.

ECONOMIC

"These new machines have a great capacity for upsetting the pres-

ent basis of industry, and of reducing the economic value of the routine factory employee to a point at which he is not worth hiring at any price...we are in for an industrial revolution of unmitigated cruelty."

—NORBERT WIENER, MATHEMATICIAN AND
EARLY COMPUTER PIONEER, 1949

Many dystopian stories of artificial intelligence run amok take place in worlds of malevolent machine superintelligences, from *The Matrix* to *The Terminator*. While technological singularity is possible, there are subtler economic concerns that don't depend on the bleak collapse of civilization. We're already charging headlong into a technocracy, with a handful of elites with know-how and an ownership stake in the machines responsible for most productivity gains. Outside of hedge funds, most new billionaires, millionaires, and top-end professionals of the past two decades have come from the ranks of the technology sector.

While productivity has relentlessly ticked upward, average employee salaries remain flat for most, but not for those at the top who either own or deeply understand the new technologies and shifting economy. As the physical labor of farming became automated, the number of people working in agriculture decreased. As the repetitive yet technical know-how of manufacturing continues to become more automated, the number of people required to do that work is also decreasing. There's no reason to believe this trend won't continue, as AI automation for cognitive, professional, and creative work is increasingly automated. What economic sector will be left for humans to move into? While there's some hope that education for new jobs will be our saving grace, it's worth noting that the education ladder is really a pyramid. There is a limit on how many data scientists and executives the world needs, and it's far less than the current demand for truck drivers. If even only half of the four million US transportation jobs become fully automated, there won't be two million other jobs available.

We need to consider the possibility that there just won't be enough jobs to go around. Economists and policy makers have floated a few options, from wealth distribution to universal basic income, based on taxing any productivity gains made by AI. Each possibility opens up a myriad more questions. How we deal with this potentiality is far beyond the scope of this book, but it's a discussion our society seems woefully unprepared to have.

PHILOSOPHICAL AND ETHICAL

In the Star Trek oeuvre, the Kobayashi Maru is a test of how Starfleet officers act in a no-win scenario. Humans rarely run into situations that lack any viable options, and the concept seemed so remote and philosophical that audiences applauded when Captain Kirk found a way to beat it. But we're reaching a point where, in a world that lets computers make decisions, humans will have to preprogram what the computer should do.

The 2004 film *I, Robot*—loosely inspired by an Isaac Asimov book of the same name—envisions a future society where humans and AI-based androids live and work together. In the film, the protagonist, Spooner, is primarily motivated by a subplot involving a car wreck and a little girl. Two vehicles slid off of a bridge: one car contained the child, and the other contained the adult Spooner. A passing android could only save one, and it chose the adult because it calculated that he had a slightly higher chance of survival. The robot pulled Spooner from the wreckage as he could only watch in horror as the child sank into the deep. Most humans, when confronted with the choice, however, would have gone for the child. We can't assume that AI will make the same decisions that humans will. For all the seeming magic of AI, it still lacks basic common sense and compassion and, therefore, must be taught by humans what is right in advance.

Furthermore, as the cognitive power of machines grow, we'll reach a stage where our own moral code must come into question. Does an android that looks, talks, and acts like a human deserve any rights at all? This dark question was grappled with in the film *A.I.*, based on short stories about an android child programmed to love (or at least convincingly simulate it) and how he and the world interact with each other. Some treated the android as a human, projecting their values of human children onto it, while others believed that artificial meant no rights at all, despite the machine's protests.

If this story seems fantastical, consider MIT's project, Quest, which builds increasingly sophisticated neural networks, like modeling a human child who learns over time. Other projects are unlocking how the brain turns raw signals into senses, perceptions, conceptions, and potentially consciousness itself. Is there an ethical consideration of a completely simulated human brain, one that acts entirely human in every way?

While these sorts of philosophical and literary situations have puzzled humanity for generations, we are the generation that has to have real answers for them. And those decisions will have real consequences for generations to come.

THE AI SUMMER

Unlike some other emerging technologies, like blockchain, whether companies adopt AI or not is not really a choice. AI is inevitable, because the economic incentives of artificial minds are just too strong to stop. But the good news is it's early enough that we can choose how to react.

While the downsides of AI may seem dark, there are many positives we shouldn't overlook. If we can get the social factors figured out, we may realize the dream of the industrial revolution: more free time for all of us to pursue personal interests. Who needs a forty-hour work week, anyway? I'd love to spend more time with my family while letting my robotic mower cut the lawn or my AI assistant answer emails. We could completely automate farming while reducing dependence on chemicals and streamline the military while reducing human casualties. We could automate the manufacturing of products onshore while reducing transportation costs and thus carbon emissions—hopefully while profit-sharing the output generated by a network of machines that require little human intervention. Unlike von Kempelen's machine, we can finally take the man out of the Turk. The dream of a postscarcity society is possible for the first time in history.

FURTHER READING

- *The Master Algorithm: How the Quest for the Ultimate Learning Machine Will Remake Our World.* Pedro Domingos
- *The Deep Learning Revolution: Artificial Intelligence Meets Human Intelligence.* Terrence J. Sejnowski
- *Superintelligence: Paths, Dangers, Strategies.* Nick Bostrom
- *Humans Need Not Apply: A Guide to Wealth and Work in the Age of Artificial Intelligence.* Jerry Kaplan

- *Rise of the Robots: Technology and the Threat of a Jobless Future.* Martin Ford
- *On Intelligence.* Jeff Hawkins

3

EXTENDED REALITY

THE METAVERSE

On March 16, 2020, in response to the COVID-19 pandemic, the state of Oregon joined other US states and countries by instituting a stay-in-place order. Offices were closed, along with my kid's school and the park across the street from my town house. *My* family was safe, thankfully, with plenty of food and toilet paper, but before we knew it, the walls started closing in. Like much of the world, we went stir crazy, desperate to leave. But the Redmond household had a secret, cutting-edge, savior technology: virtual reality (*VR*). It became a respite, a transport beyond the four walls, and offered us a modicum of social contact when we were robbed of so much human presence.

VR teleports our minds into a world unreal. Unlike video games or video chat, which trigger one or two neurotransmitters, VR has been shown to trigger more, enabling a sense of brain-body presence that other technologies can't match. Researchers call this "deep embodiment," which can increase feelings of connection and fight the isolation inherent in mass quarantine. Its power is difficult to describe—much like reading about the taste of food or the sound of music. Language doesn't do it justice. But here's a small example of how it can profoundly impact our everyday reality.

During the COVID-19 pandemic, I joined others in the great transition to working from home. Being a team of mostly software engineers, our day-to-day work could be comfortably executed remotely. We had long ago adopted real-time chat apps like Slack and kept track of our work in a digital Kanban board with computer source code on GitHub. The only major change to our work was that our daily standup meetings, Agile Scrum, were done remotely through Zoom video chats. But no matter how many tools we adopted, something was missing. We completed our work, but the sense of camaraderie faded. That ineffable sense of companionship—of being around one another, sparking unstructured dialogs and socializing in person—was lost in sanitized video chats. It was better than email but still a disappointing simulacrum of true presence. So, being a team of professional problem solvers, we hunted for a solution. We first tried keeping an ever-present video chat open, but the endless apologies of background noise and a burning need to explain each time you stepped away from the keyboard quickly grew tedious. We structured team lunches, which were nice, but a screen covered with a dozen closeups of ravenous maws can kill an appetite. My team is mostly teetotalers, so happy hour was out. What were we to do?

By pure dumb luck, I had happened to buy top-of-the-line VR goggles for my entire team a month prior to the pandemic as a reward for delivering an unrelated project. So we looked to VR as a potential solution. After several false starts, we finally found a service called *Immersed VR*. This app allowed us to connect to one another in a virtual room, with an avatar representing each team member. We could show one another our monitors, the virtual version of leaning over your coworker's shoulder to collaborate on a problem.

Best of all was the virtual whiteboard. Robbed of a physical office space, this allowed us to stand around a whiteboard, draw out concepts, and speak to it. Craziest of all, since we connected to our physical laptops in a virtual world, we were still able to answer emails and attend the odd Zoom meeting inside of VR. We looked strange, of course. Those non-VR video chat users would see me and my team in all of our webcam glory sporting bulky Oculus Quest goggles on our faces, wondering what in the hell we were up to. The future always looks a bit geeky before it becomes cool. From our vantage point, we were taking a Zoom call from space.

My coworker on video chat in one of many floating virtual desktops.

Regular workdays were only the beginning. We began attending conferences in virtual reality. Saving the cost and time of travel or nights alone in janky hotels, we were able to walk around a conference space, attend more lectures together, and comment on them live. We got to listen to world-class speakers and didn't miss out on the after-hours water-cooler talk. The flexibility also allowed us to hop in and out of other meetings or conflicts. And in one case, we hopped between two simultaneous conferences. We wandered around V-Market, the world's largest virtual bazaar. We were stuck at home, and yet we were everywhere.

For those of us over the age of thirty, VR has been an exclusive vehicle of science fiction, starting with Neal Stephenson's 1992 novel *Snow Crash*'s VR Metaverse, more richly realized than William Gibson's "cyberspace," coined a decade prior. Shortly after, from *Lawnmower Man* to *Johnny Mnemonic*, from *The Matrix* to "the Oasis," we could only dream of experiencing such virtual worlds. This began to change in 2012 after two major launches: Oculus Rift and Google Glass. Both products created real-time computer-generated worlds, bookending the ranges of extended reality (XR), virtual reality (VR), and augmented reality (AR). True, better options existed in labs and expensive niche business cases, but these captured the public imagination due to their quality and consumer accessibility.

In the virtual reality camp, a nineteen-year-old named Palmer Luckey resurrected the old dream of virtual reality. Luckey's obsession consumed his life; he lived in a trailer and worked at a research group while he gained enough knowledge to put together a prototype of the highest-quality VR headset built from cheap IoT components. He funded his work with a motley crew of 3D luminaries and crowdfunding. This effort was quickly followed by several major corporations from Samsung to HTC, and in 2014, Facebook bought Oculus for $2 billion, officially kicking off the VR revolution we're now in the early stages of building.

Around the same time, Google's augmented reality eyepiece called Glass was making waves. At $1,500 a pop, it failed to penetrate the market like Oculus, but it gave the world a glimpse of what could be achieved in an eyeglass-sized form factor. In 2014, I navigated China with only Google Glass and a visual translation application, lacking any knowledge of the Chinese language or logograms. The Glass app would instantly translate any sign I looked at and replace the text with English words (this feature is now built into Google Translate, but at the time, it was daunting to say the least).

In nine short years, we've moved from VR and AR being presumed as lost causes to becoming the most exciting general purpose technologies in play. How and why are fascinating questions. But before we answer them, let's dig into what makes up this landscape of extended reality and what we can do with it.

THE MANY WAYS TO EXTEND REALITY

The earliest VR, MR, and AR were able to penetrate the mass market only because they hijacked an existing product that everyone already owned: smartphones. Samsung Gear and Google Cardboard were cheap, on-face housing units that mounted a smartphone's screen close to the user's eyes. Leveraging some clever optical lenses, software, and built-in components like the phone's accelerometer, a passable VR experience has been within anyone's reach for a few years. Other early devices leveraged existing platforms, like Sony's use of the PlayStation console to drive their first VR games. As for MR/AR, iOS has ARKit, and Android supports ARCore. Both are hardware/software kits that allow any developer to create mobile AR experiences with little expertise. Doubling down on MR, the newest Apple products of 2020 support LIDAR, the same world-scanning technology that makes self-driving cars possible. These core libraries are paired with 3D development engines like Unity or Unreal, letting designers import 3D assets and control what a user experiences when, for example, they point their phone's camera at a certain poster or QR code. If programming is too heavy of a lift, there are drag-and-click development environments like *Adobe Aero*, democratizing MR/AR design and placing it within anyone's reach.

Let's click down into VR, AR, and MR, and find out what they have in common and what differences exist between these technologies.

VIRTUAL REALITY

"You empathize with her in a deeper way," Chris Milk told a TED audience. He was the cocreator of an "empathy machine," an experiment designed to bring world leaders to a young girl in a Syrian refugee camp with a deep sense of presence. It's a heart-wrenching video made real by way of VR. Considering that the attendees at Davos, TED, and the UN make decisions that affect millions of lives, this simple act of empathy generated real-world consequences.

What is VR? It's a simulated experience, but that meaning requires a lot of caveats and technologies. Is it limited to the visual and sound? What about other sensations like smell or warmth? Ever since *The Matrix*, questioning the limits of our senses and the nature that of reality has become a mainstream issue.

The fundamentals are simple enough in our modern age. Smartphones created the right conditions where sensors were cheap enough that a teenager could build a decent VR headset with the help of some internet friends. The basic components of a virtual reality headset are high definition video displays, ocular lenses, accelerometers, and software or hardware that turns a scene based on the direction where a user is looking. It renders a slight shift for each eye (called *parallax*), to create the illusion of 3D depth. This simple build can provide three degrees of freedom (or *3DoF*, verbalized as "three-dof").

The more components leveraged, the greater the experience. In the simplest VR, a smartphone and Google Cardboard, you're limited to looking around a 3D scene in a seated position. Positional and standing VR, like Sony PlayStation VR or the Oculus Quest, are called 6DoF. An incremental improvement to 6DoF is called *room-scale VR* and allows you to move around a fixed space. This requires beacons called *base stations* (or lighthouses) that track exactly where and how the headset is oriented within the space, or a space-mapping technology called SLAM (simultaneous localization and mapping). Many game makers like Zero Latency circumvent space constraints by creatively guiding you around in circles and zigzags, so you get the sensation of a larger scale without running into any walls. *The Void* plays a similar trick but within a small labyrinth.

To further increase the sense of real presence, wider screens and optics provide a greater field of view (*FoV*). A narrow FoV like ninety degrees will feel as though you're looking through a tunnel. A wider FoV, around 180°, will start to seem more life-like, as you can no longer see the sides of the headset. Generally, FoV is measured horizontally, but some measure diagonally and vertically as well.

Another improvement in VR is reduced pixilation, or screen-door effect (*SDE*). Imagine pressing your face close to an old computer monitor. The crisp scene you see from a distance starts to look like a matrix of pixels. Since VR screens are so close to your eyes, the screens need higher density to avoid showing individual pixels. In other words, the pixels physically need to be smaller. Looking at a scene that's only a few hundred pixels gives the sensation of looking at it through a screen door, hence the name. Newer 4K- and even 8K-high-definition headsets are increasingly available, reducing the sensation of SDE. A more advanced trick called *foveated rendering* actually tracks the movement of your pupils and renders exactly where you're looking in higher fidelity, allowing for a more realistic experience while requiring a minimum of computing power. The VRgineers headset *XTAL* uses both 4K and foveated rendering to generate an experience so lifelike that I can easily read small pale text inside of a VR cockpit.

There are dozens of improvements coming to VR over the next decade. Improved fidelity, reduced distortions and screen issues, resolution of comfort issues like size and weight, better mobility with standalone headsets (not plugged into a computer), and of course reduced price and more applications. Soon VR will look as real as the actual world through goggles as comfortable as sunglasses.

AUGMENTED REALITY

When Google Glass first came out, I wore that uncomfortable prism on my face for an entire year and wrote two books about the revolution that never happened. Glass was an augmented reality device intended to change everything, like the Segway or the Krummlauf. But all is not lost: it's finding a new life for industrial and medical applications.

Augmented reality is distinct from virtual reality and, in many ways, is its polar opposite. Where the purpose of virtual reality is to generate a distinct and immersive world, augmented reality intends to add more information to the real world while you're engaged with it. In some ways, these two approaches are exemplified by two companies championing each approach. Google, creator of Glass AR, wants to be your source of information and guide in the real world. Facebook, owner of Oculus VR, wants to create a separate world comprised entirely of social fabric.

When discussing AR, I'm talking about a head-up display (HUD), where information is rendered over a scene but is not necessarily a part of it. Imagine looking at a work of art where the title and artist's name is rendered around the art, but it's not necessarily rendered at a fixed location. When augmented reality more completely intermingles with the real world, it's sometimes called *mixed reality*, which we'll cover in a moment.

From Vuzix for biking to Form Swim Goggles, there are plenty of players in this space, each designed to provide information to users in a hands-free way. Fighter pilots have used HUDs for decades. The technology behind HUDs is relatively straightforward and largely an exercise in small, low-power sensors and extremely high-definition micro displays.

AR extends beyond face-mounted systems to include smartphone apps through a camera, leveraging built-in SDKs, such as Android ARCore and iOS ARKit. This same technology can be placed on transparent displays in the front glass or smart display cases, in mirrors for help in applying makeup or exercising, and in cars that project speed and directions onto the driver's side glass.

All of the components in this simpler form of AR are important for the evolution of the more complex form of AR called mixed reality (MR). The current primary tool of both AR and MR is the smartphone. The important task over the next few years will be leveraging smartphone AR/MR capabilities in preparation for wearable MR. AR built into smartphones created an army of developers and companies comfortable with developing these sorts of apps and testing the user experiences in the real world before the tech is at its final stages. AR HUDs continue to push the boundaries of technologies that can be packed into smaller form factors that render high-definition images near a user's eye, while extending battery life and reducing weight and heat. Focals by North, purchased by Google in mid-2020, are pushing for smart glasses that are stylish, while patents and rumors have it that Apple is working on an AR glasses project of its own. The ultimate goal is to reduce or remove the need to carry smartphones, with virtual devices that you only see and hear.

MIXED REALITY

The difference between augmented reality and mixed reality can be confusing, especially since AR is often used as a catch-all term that includes MR. In general, AR includes any digital augmentation to actual reality. What makes MR a distinct sub-discipline is that MR technology requires a detailed awareness of the surrounding world. It's not enough to have a video screen hovering in the air like a head-up display—in MR, that video screen appears mounted to a particular wall. It mixes together the real world and the virtual world. The goal of MR is to create the sensation that digital objects are indistinguishable from actual reality.

This mixing requires similar components of VR, plus more. MR is significantly more difficult to achieve and is, in some ways, the holy grail of XR. MR has similar components to VR, such as a wide field of view lest digital objects disappear as they move outside the display, 6DoF so walking toward digital objects makes them realistically appear to be bigger, and 3D rendering of objects close to someone's eye, leveraging forced perspective. This difference is what makes MR technology so interesting and difficult.

Mapping digital objects onto the real world in a believable way requires the headset to be able to superimpose digital objects onto a view of the real world. This can be achieved by either overlaying a 3D digital rendering on a transparent display or by rendering onto a video that's captured by a camera. If it's the latter, the MR system needs one or more cameras, and can render the scene on a single screen or render it twice with a parallax effect on each screen per eye.

The second major need is highly detailed SLAM, allowing an MR device to orient itself in space. This is much harder than it sounds, employing multiple cameras, gyroscopes, accelerometers, compasses, and more exotic components from radar to lasers. It's important for the device to know a precise position in space because it must superimpose a 3D digital object in relation to surrounding real objects with a high degree of fidelity. Otherwise, the user will experience an uncanny sensation of drift. Also, if you place a digital ball on a table and turn around, when you turn back, you expect the ball to still be there. If you sit on the couch, you expect the ball to appear to be sitting on the table, not slide inside a chairback or hover an inch above. Beyond object permanence (something even babies expect of the world), the device needs to know where you have moved in space. The first step in rendering the digital scene correctly is for the MR device to have an internal sense of the real 3D world, which requires endless scans, updates, and fusion of various sensors like LIDAR.

The next difficulty for mixed reality is *occlusion*, or blocking virtual objects with real ones. If you've seen sports on television, you're familiar with the concept. When an American footballer runs across a virtual line on a field, the person covers the line; the line is not drawn over the person. In other words, Drew Brees's body occludes the virtual ten-yard line. Straight lines on a generally green field are one thing, but technology that follows a real child as they occlude the scene of a rendered virtual train set beneath a Christmas tree is something else entirely. The MR device needs to know exactly where the train should be placed in the real world (SLAM), render that scene (transparent display), and track the child as well, masking their silhouette so it appears as though they walked in front of the tracks (occlusion).

Magic Leap's light field technology, Apple's 2020 pixel projection patent, and Microsoft's vast improvements between HoloLens 1 and 2 are but a small set of cases. There's still much to do in the realm of MR, but we're making fast progress. Perfection is not necessary for mass adoption.

THE PROMISE

Google, Facebook, and Microsoft are all involved in extended reality in some fashion. While these three companies aren't the only players in the XR game, they're certainly the biggest. And the kinds of devices they support shows quite a bit about the culture and purpose of each company.

Google wants to be your guide in the real world, and their

flagship augmented reality device, Google Glass, is a pure distillation of the company's goal to "organize the world's information." While Glass was famously a consumer failure, it still exists for enterprise use, and the high price tag will come down over time, making it more accessible. Glass is a voice activated, AR HUD that projects images and video above the wearer's field of view. The goal isn't to *overlay* what the wearer sees but instead to *provide information* about what the wearer sees. Underlying all of this power is Google's vast data and AI platform. What Glass does, effectively, is bring Google into the real world. Google doesn't want to fight for your attention—they want to act as a deeply personal assistant, allowing users to more fully engage in the world.

Facebook is the polar opposite of Google in the XR space. While Google doesn't want your attention, Facebook wants all of it. They want to build a world for you, an oasis you can escape to and live in, comprised of games, work, and of course, your social fabric. Facebook's acquisition of Oculus VR was a clear evolution in this quest. Over the previous decade, Facebook has invested more heavily into scaling out virtual reality than any other company, and their gamble is starting to pay off. Oculus Quest was the first VR headset that was entirely standalone, meaning it didn't need to be plugged into an overpowered gaming computer for use, and had an increasing catalog of tools and games for work and play. Many of the technical details have been solved, and the Quest is the benchmark against which all mass-market consumer devices must distinguish themselves.

Oculus Quest is VR simple enough for kids.

Microsoft wants the middle, where you integrate the real world with virtual experiences. Like Facebook and Google, Microsoft's play aligns deeply with its corporate culture. Microsoft's HoloLens 2 in 2020 was a huge leap beyond any competing mixed reality devices and targeted Microsoft's sweet spot: the corporate market. Considering the vast opportunities for overlying digital directions onto the real world (from constructing an automobile to replacing a printer head), MR is the big fat middle that they bet will pay dividends down the road. A Forester report estimates over fourteen million US workers will wear smart glasses by 2025, and Microsoft wants to be the platform of this important market.

HoloLens 2 used in a doctor's office.

Beyond these big three representing three different approaches to extended reality, there are numerous specializations of approaches focused on various markets from mass to niche. Some (like gaming) are obvious, while others (like mental health) are not so much. But one thing is for certain: over the next decade, there is scarcely an industry that some form of XR won't touch. Here are a few.

GAMING

With VR games, you can go deep-sea diving, fly through the air as a bird, fish exotic locations, fly on a rocket to a remote planet, hunt zombies, or merely sit quietly in a calm field in the rain. Playing in a band on stage to a huge virtual audience might sound like any old rock-star game, but the rush of peering across a sea of rendered faces approaches the real thing—including small twinges of fear. I said earlier that XR was difficult to describe, but I'll try to describe it in two words: Tetris effect. XR has the potential to engage your mind and senses to the point where games and patterns become imprinted into your thoughts and ways of seeing—it drags you in. Now imagine that potential combined with the addictive power of games.

Anyone who knows the video game industry knows that the business itself is no game, with sales topping $120 billion in 2019 and growing faster than the general economy. The current renaissance in the XR space has been largely driven by video games, or at least leveraged gaming possibilities to generate excitement. Consider the Oculus Rift VR headset. It began as a crowdfunded campaign led by gamers who were hungry for the possibilities of VR. On the other end of the spectrum, Microsoft's MR HoloLens rig's first public demos heavily pushed the narrative where users could play games enmeshed in the real world. On a stage in Redmond, Washington, wearers were treated to a future where your bare hands could be guns and any room could house monsters. All you needed were the right lenses. One of my first experiences with an AR game was playing live-action *Pac-Man* with Google Glass AR, with virtual pellets and ghosts, and the city streets of Portland, Oregon, was the playing board.

As the market grows, studios are starting to release games exclusively for XR, as well as modifying old games to work in VR. An example of the former is the VR hit *Beat Saber*, where a player swings two virtual lightsabers at a field of incoming shapes, with the beat of a soundtrack. It's *Guitar Hero* for Gen Z. As for modifying older games, even game makers without a strong VR presence are getting in on the action. Nintendo updated their hit *Super Mario Odyssey* to work in VR mode, showing just how easy it can be to retrofit existing 3D games and literally put yourself in the game. In early 2020, Valve's *Half-Life: Alyx* was released, and as far as VR games go, it was a hit. It ran 43,000 concurrent players and over 300,000 watchers on its first day.

TOURISM AND EDUCATION

Tourism is fun and educational, a major global industry, and a serious pastime of my family. There are well-studied benefits of global travel beyond family bonding, fostering cross-cultural literacy and parasocial connections. Sadly, there are major barriers for anyone who wishes to experience another land. The first is spatial; the second is temporal.

Merely getting to another country is beyond many people's abilities—be it the insurmountable cost of travel, legal restrictions, or health. A straightforward benefit of virtual reality is that it allows anyone in any situation to experience the feeling of being present in any number of real locations. Ascape is one of many companies that provide 360-degree virtual experiences of famous locations from Serengeti safaris to helicopter tours of Dubai. But there's a practical sort of travel as well. YouVisit is a company that provides prospective students with virtual campus visits, while the game *Boulevard* lets users tour priceless works of art through world-famous museums, all without fighting crowds or leaving the couch. Maybe physical or financial restrictions don't apply to you, but there's a VR app for you as well. Before booking a hotel or resort, there are increasing 360-degree video VRs available, letting you virtually tour a hotel before you charge a credit card.

The temporal barrier to travel is a bit different but can open up travel opportunities beyond virtual. During their first week of homeschool, my children visited the Great Pyramid of Giza. My daughter excitedly stood at the foot of the Sphinx and remarked about its missing nose. Her four-year-old brother liked to stand under the camels. Going beyond any classroom or lesson plan, ancient Egypt became real for them in a way no amount of video or description could match. It sparked a deep interest in them that lasted for months, and potentially a lifetime.

Beyond virtual reality, mixed reality is also a valuable tool for extending real-life experiences into the past. We can save historical sites like Pompeii from being littered with informational plaques and navigation signs, offering augmented digital information instead. It's a new form of travel, where ancient sites may be experienced as they were, giving us a closer connection to our ancestors beyond dusty monuments and bones.

VIRTUAL SHOPPING AND AUGMENTED RETAIL

One of the best things about virtual, mixed, and augmented reality is the combination of the digital and physical world to make the unactual feel actual. This is fine when you're enmeshed in a world of virtual objects, but to quote Woody Allen, "I hate reality, but it is still the only place where I can get a decent steak." Sometimes you need real things. Sometimes you need to go shopping.

Obsess is a shopping platform where you can wander through an interactive brand experience set up as a virtual retail store, complete with the same jazzy soundtrack you'd hear in a physical space. As you navigate the floor, you can pick up products, get information on them, find sales, and buy them right from the headset. Amazon has done something similar, but instead of being confined to a virtual store, shoppers can instead experience a virtual hot air balloon ride. Blurring the line further between physical and digital retail, many retailers now augment their physical retail spaces with AR applications for more information or individualized pricing.

IKEA launched one of the first mainstream MR shopping applications, where users could scan any of the rooms in their home with a smartphone and drop virtual furniture into the space. It's easier than shuffling chairs around the room, and it lets you walk around the space to get a feel for how, for example, a TV stand would look before clicking the "buy" button. Not to be outdone, Nike launched a comparable MR experience for footwear so you can see how those Jordan's will look with your jeans.

Augmented retail is not just for shopping but also for brand enhancement. Regal Cinemas has an AR app that allows you to experience short digital storylines, making movie posters come alive. In the early days of the millennium's second decade, most of these experiences required looking through a smartphone. But it's only a short matter of time before this capability will be ever-present through wearable AR glasses.

XR INDUSTRIAL DEVELOPMENT

In the realm of XR industrial development, few companies have been on the forefront like SpaceX. In a demo from 2013, Elon Musk showed off a virtual rapid prototyping toolchain straight out of *Iron Man*. Using hand gestures, he manipulated a 3D virtual object and sent it to a 3D printer, where it was fabricated in titanium. Though it was largely a demonstration, it gave an early glimpse of what modern 3D industrial design could be. Fast-forward a couple of years, and Ford Motor Company's immersive vehicle laboratory started to demo how they design cars in VR. Not only can engineers create and modify parts, but designers can experience how a driver or passenger will experience and interact with designs not yet created. This helps create better user-centric designs and can save millions in physical modeling costs, all for the price of a VR headset and some software. With HoloLens, MR designers can now modify existing products by superimposing digital edits over real cars.

We live in a global world where experts abound. As work becomes more distributed, extended reality is a simple solution to ensure collaboration is not lost when physical proximity goes away. As industrial designers increasingly use XR to develop products, they use that same technology to interact with colleagues across the world in real-time in a personal way. For example, Facebook's Oculus Go is a simple and inexpensive way for teams to meet in a virtual space and collaborate.

There are countless more industrial use cases for XR. VR environments can help airline pilots train without ever leaving the ground, MR goggles can overlay technical information onto physical machines to aid in repairs, and AR can help warehouse workers navigate a large warehouse. Enterprise XR may be the first exposure many people have before it creeps into our personal lives.

HEALTHCARE AND MENTAL HEALTH

When a friend contacted me to join him in a healthcare augmented reality startup, I thought he was crazy. Not because of the technology—I had just completed a book on Google Glass AR and worn Glass for a solid year—but because healthcare is a notoriously tricky industry to serve. It's a labyrinth of competing concerns, standards, and regulations, with dire consequences if anything goes wrong. But we dove in and built a product that allowed a doctor to perform hand surgery while another physician proctored from a remote location. The surgeon streamed a live feed using a high-quality loop camera, while the consulting doctor monitored the situation on a browser and spoke with the surgeon in real time. He was able to draw on the video and share the augmentation back to the performing surgeon with visible notes like "Cut here" appearing on a screen that hovered above the surgeon's field of view on Glass. Normally, proctoring takes days of planning and travel for just a fifteen-minute consult. This single project allowed the best specialists in the world to consult with dozens of cohorts in a day rather than just one or two a month and at a fraction of the cost. Although our little startup couldn't cut through the complexities of the industry, similar tools live on, such as *Proximie*. In the MR space, there's Augmedics for surgical guidance, which is like seeing through a body with all of the anatomy labeled.

Beyond aiding collaborations and surgeries, VR has great value in data visualization, medical student education, and training with new equipment or techniques that enhance patient treatment and outcomes. XR can also reduce waste, such as by leveraging virtual cadavers for anatomy labs.

DTx, or digital therapeutics, is a catchall for technologies in support of therapies. One of the most interesting DTx is for mental health. There are VR projects dedicated to mindfulness and meditation, as well as cultivating a sense of relief and connection to help you take care of your mental health. One of the top apps across VR stores is *Guided Meditation VR*, which places the user in their choice of calm environments (I like rain) and can guide you through breathing and mindfulness exercises.

For those who need more intensive therapy, VR is a new vehicle for dealing with certain phobias and traumas through safe immersion therapy. Say you're afraid of heights and want to work on reducing that fear. Samsung has a VR experience called *#BeFearless*, designed to help you virtually work your way up to greater heights. Unlike real exposure therapy, you can take your headset off if the feeling gets too intense. Outside of the do-it-yourself mental health options, there are VR tools designed to support professional psychologists in their practices. *Psious* is a kit that includes VR and biofeedback to support expert guided methods, like cognitive behavioral therapy.

Imagine starting a big web business in the mid-1990s. That's where we are on the curve. From games to travel to new shopping experiences and healthcare, the benefits of XR are just being discovered. The next generation of Amazons and Googles will start this decade.

MORE TECH TO EXTEND XR

We began this chapter by focusing on the composition of XR, followed by promises and real-world applications. Now, let's peek at the more recent technologies being developed to expand the experience of XR.

The lobby of Xerox PARC (Palo Alto Research Center) is a mini museum of technology history. Nestled behind a thick glass case, inspired by what was rightfully dubbed "the mother of all demos," lay one of the first graphical user interfaces and mouse. Before these inventions, the most advanced standard interface next to computers was text on CRT monitors with keyboards and, not long before that, punch cards. This simple alteration of how people interacted with computers inspired both Apple and Microsoft to launch the PC revolution in the early 1980s. Today, we're still in the early, awkward teen phase of the XR revolution. The best is yet to come thanks to a bevy of technologies occupying research labs in 2020.

CONTACT LENSES

The 2006 near-future science-fiction novel *Rainbows End* predicted a world where everyone will wear mixed reality contact lenses. The author, Vernor Vinge, considered that in a world where computer imagery continually overlays the real world with a virtual one, all advertising would be virtual, as well as building facades and even computing devices. Why hold a smartphone when you can project any video you need? Why bother painting a building for your bakery and hanging a sign, when any gray concrete enclosure can simply project a beautiful French design? This sci-fi world may be closer than you think.

In 2014, Google announced a project to measure glucose levels for diabetics by way of smart contact lenses. Although the project was eventually scrapped in 2018, the experiment proved that electronics had shrunk to a small enough size where they could be housed in contact lenses. Fast-forward six years, and the company Mojo Vision successfully demonstrated a smart contact lens prototype that included a 14,000-ppi (points per inch) display, complete with eye tracking, image stabilization, and other sensors. The lens is capable of projecting a monochromatic image directly onto the eye, thanks to its high-density display (though in 2020 it was powered by an external battery pack). It's still a few years from production, but it's a big start.

So, how can we control these lenses? Shortly before Mojo Vision debuted, scientists at the University of San Diego created contact lenses capable of zooming in on a scene, activated by various blinking rates. Beyond that, hands-free gestures are already available with many XR devices. By the time AR and MR contacts exist, we'll be accustomed to swiping virtual objects in the air. It makes sense to translate this user experience to AR/MR contact lenses.

FULLY BODY HAPTICS

It's difficult to discuss the future of VR without citing the novel *Ready Player One*. This book, which became a Steven Spielberg movie/nostalgia commercial, helped promote a world where everyone lives much of their day in a hyperrealistic, massively multiplayer virtual reality world. Beyond VR goggles, it also sported accessories to make the experience more realistic—namely, haptic suits and omnidirectional treadmills.

Two things missing in today's immersive world of VR are the lack of tactile response, and the fact that—while a VR world may be unbounded—you'll eventually run into a wall or, worse, a metal bench at shin height.

Initially, the best solution to these movement problems was a large warehouse size location, such as Zero Latency in Las Vegas. But now, the easiest solution is omnidirectional treadmills. In early 2020, *KatVR* was ahead of the pack as a small, bowl-like treadmill with a pivoting harness that holds your body in place; you even get special slippery shoes. This allows you to walk, run, or sprint in any direction without physically changing location, but you're given the virtual feedback of having moved. These treadmills avoid the problem of real obstacles when such things don't exist in the VR world you're inhabiting. The experience is bizarre at first but quickly begins to feel like natural locomotion, all for under $1,000. But locomotion with virtual position tracking is only half of the solution.

The endgame of virtual reality is full haptic feedback. A handful of companies are working on and selling wearables that synthesize a tactile sensation and force feedback from VR objects. Rather than your hand going right through a virtual baseball bat, force feedback mechanically stops your hand from penetrating the surface. When a ball hits your virtual bat, a tactile push and small vibration can inform your brain that something was struck. The company at the forefront of the haptic revolution is *HaptX*, which manufactures feedback gloves. I tried them once while hovering—disembodied and godlike—over a VR Lilliputian farm in space. The small plot held a cartoonish barn, tractor, crops, and surrounding phenomena. Suspended on celestial skyhooks were clouds. I squeezed a fluffy white one, evoking the sensation of a rubbery wad of paper. Like a power switch, it activated the rain, making the cloud turn gray and thundery. I ran my hands underneath and felt the tingling sensation of individual drops, followed by an electrical surge of a cartoon lightning bolt. It's one thing to see virtual realities that aren't there; it's another experience altogether to feel them.

However, we interact with the world using more than our hands. Teslasuit is a full body haptic suit that covers a wearer's body, arms, and legs. Not only does it provide touch sensations via haptics, it also supports a high degree of motion capture, helpful for creating a more realistic VR model. It also senses your heart rate, allowing an XR environment to react to your motions for training and biofeedback.

BRAIN MACHINE INTERFACE

In the Japanese anime show *Sword Art Online*, people play in a future where VR has been perfected so completely that bulky contraptions such as treadmills, haptic suits, goggles, and headphones aren't necessary to experience a fully immersive virtual world. In the show, they call this *deep dive*, where signals from the brain are captured directly as output, and a virtual world is sent directly to the brain as electrical signals. This includes every sensation, from sight, sound, and touch to even smell and taste. While it's easy to write off these ideas as hard science fiction, there are a few companies working toward versions of this goal.

In the world of reading neural signals, the darling of CES 2020 was BrainCo. They demoed an EEG headband that could capture thousands of brainwave signals. Leveraging hydrogel sensors and AI, this headband could read neural signals with a higher degree of accuracy than some laboratory devices. What makes BrainCo interesting is that, with minimal training, you can learn how to connect to a virtual world within in a couple minutes using only your mind.

Reading neurons is one thing, but writing them? That requires a more invasive option. One of Elon Musk's stranger ventures is his founding of a neurotechnology company called Neuralink. Neuralink creates a high-bandwidth brain machine interface (BMI) focused on allowing two-way communication between a human brain and a computer. Neuralink's procedure implants a neural lace that connects directly to neurons, which in turn connects to devices capable of both transmitting neural signals and accepting inputs to turn on neurons. Such integrations start closing in on transhumanism, or cyborgs, and bring up all sorts of existential questions on what it means to be human or machine. But whatever it is, it's also one step closer to fully experiencing and integrating wetware and software, or in other words, the actual and the virtual.

PROBLEMS IN XR

Despite its clear advantages, this decade presents a fair number of blockers to mass market adoption of extended reality. As someone who wore Google Glass on his face for an entire year—a so-called Glasshole—I can assure you that social acceptance of ever-present augmented reality has a long way to go. Legal questions around privacy, social contracts, and ownership of digital assets are still wide open. All of these answers must also deal with the fact the technology is in constant flux. So, how do we support artistic expression without adopting an overly laissez-faire attitude toward the very real possibility of psychological torment? Technology is rarely inherently good or evil; it just makes everything more powerful. XR may be one of the most powerful technologies in terms of social effect since the invention of propaganda. The decisions we make this decade will affect humanity for a century to come.

GETTING PEOPLE TO CARE

When IMAX opened a VR cinema in LA in 2017 to much fanfare from the community, many enthusiasts thought, "Finally, VR is becoming mainstream!" A year later, IMAX quietly closed the doors. With XR, you're asking people to operate in a new world in a very different way. When confronted by eager business owners dreaming of world domination, programmers often joke that scaling a product is not the problem—it's getting people to care. This is as true of XR as it is of any other tech. For instance, Oculus intended to have one billion users by 2019 but delivered a mere 300,000 units.

There's a growing library of XR content and applications, but user interest, organizational process changes, and content need to move in lockstep to increase adoption rates. If one of these key components are missing, it can stall adoption, as users or content creators merely give up on waiting for others to keep pace.

ADDICTION

Modern social gatherings are characterized by people being together physically but separated mentally as they stare at their smartphones in emotional isolation. VR can exacerbate this. *Ready Player One* is as much a dystopian alarm bell as it is a fun future romp. The main characters have friends, play, and go to school online. But people also use the virtual world in the novel, "the Oasis," for escapism, and ignore the real-world plights of climate collapse, mass poverty, and legal indentured servitude. It's not hard to imagine equivalent social disruptions in a near future, where total escapism is a mere headset away. Gaming addiction is a recognized illness by the World Health Organization. China, out of concern that social networks were becoming too addictive, began requiring that major social media companies enact antiaddiction algorithms into their systems. While the best solutions are hopefully not as heavy-handed as government edicts, this analog does illustrate that society and technology have roles to play in tamping widespread social excesses. We can't hope for a spontaneous outbreak of personal responsibility at scale.

In addition to virtual worlds and video games, virtual pornography is a growing industry. Porn has historically been a major attendee and sometimes driver of many technology adoptions, from photography to VHS—and XR is no different. While remote sexuality can be enhanced by way of VR for many positive uses, such as for troops stationed abroad, the heightened experience of virtual sex is already becoming a documented addiction in its own right. This and other forms of XR addiction will require their own methods of treatment.

TRAUMA AND TORTURE

Exposure therapy by way of VR is useful to help individuals overcome phobias through safe, controlled experiences,[4] but what of the inverse? In-game trauma that feels real may cause negative side effects that extend into the real world. A virtual form of shellshock by prolonged exposure to war games can cause PTSD symptoms similar to the real thing. You may believe the risks to be slight, considering the controlled environment, but panic attacks and even heart attacks are real risks. Being scared to death, while rare, is a real phenomenon. However, studies have shown that in extreme situations, users won't necessarily think to throw off a VR headset. When confronted with strong feelings, your lizard brain takes hold. This is the evil twin of the mental health benefits provided by XR.

Without rules of conduct, nothing stops someone from selling realistic virtual sessions where real people can inflict trauma on realistic virtual avatars. We're now in the age where people can act out virtual fantasies that would make Marquis de Sade blush. Gang and terrorist organizations can recruit, condition, and brain wash en masse with no risk to their core team, and the risks extend beyond criminal groups. There are no current laws or treaties to stop military or police use of VR in prolonged and

---

4   Debra Boeldt, et. al., "Using Virtual Reality Exposure Therapy to Enhance Treatment
     of Anxiety Disorders: Identifying Areas of Clinical Adoption and Potential Obstacles,"
     Frontiers in Psychiatry 10: 773. 2019. DOI:10.3389/fpsyt.2019.00773.

controlled situations. They can simulate torture with extreme virtual, audio, and haptic stimulations. We need to work toward a globally aligned philosophy as to where the line should be drawn between actual and virtual torture, especially in a world where the distinctions between actual and virtual reality are becoming increasingly blurred.

SECURITY

In May of 2020, rap artist Travis Scott held a virtual concert in the game *Fortnite*, which was attended by twelve million viewers. Compare that to Woodstock, the concert that defined a generation in 1969, which hosted only 400,000. Scott's avatar stood five hundred feet tall and was viewable by anyone on a remote in-game island, complete with a massive stage, backup dancers, and pyrotechnics. Between the beats of his newly launched songs, the entire sky lit up, spaceships flew, and stars fell from the sky. It was a sight to behold. Larger than life, his avatar wore virtual signature Nike Air Jordan 1s, each one larger than a building. For players of the game, they too could buy virtual Nikes for their Fortnite avatars, costing them between $13 and $20. These digital objects have value for the owners—they're not only cultural cache, but have monetary value. In many platforms, they can be bought and sold for real money.

As we spend more time in VR, AR, and MR, virtual assets will take on new meaning. What do we do if someone steals these assets? What legal liability does a game platform have? Currently, these issues are up to the game maker's terms of service, but eventually, society will demand some consistency. Imagine generations hence, and your great-grandchildren losing their grandmother's digital photo collection in an act of digital vandalism or their father's one-of-a-kind championship gaming avatar being stolen by hackers.

One place where ownership and XR intersect is called *Decentraland*. It's a blockchain platform where people can own virtual plots of land, dress their avatars in virtual clothing, and resell assets in an open marketplace. In addition to these sorts of technical solutions, we'll need broader digital rights akin to physical ownership with more explicit legal and criminal governance, as well as methods of making owners whole in the case of loss and theft (likely including insurance). We may someday see digital restitution for those who—through historical inequities like the digital divide between rich and poor—were unable to take part in early digital economies.

TECHNICAL

A handful of technical challenges for AR, MR, and VR remain. Some are unique to a given variety, and some are similar across all types of XR. For example, all of them work better with wider angle of view (AoV), field of view (FoV), and degrees of freedom (DoF). Batteries are another technical challenge as headsets become untethered from larger computing devices. *The Void* VR game has no wires, but a heavy backpack containing a full-scale microcomputer is lowered down and strapped onto a player's back. The standalone Oculus Quest can last a couple of hours, but with a variety of external battery packs, it can extend to indefinite use, as long as the user doesn't mind swapping batteries every few hours.

The hardware and software necessary to alleviate physical discomforts are improving incrementally. Motion sickness is one problem, caused by a disparity between brain expectations versus the inner ear. Another is cyber sickness, caused by factors like field of view, interpupillary distance, latency, weight, and head pressure. These details, along with price, will improve gradually as components become lighter, cheaper, and more precise. Form factors are also improving. I'm personally a big fan of Panasonic's 4K HDR VR goggles, which evoke a future steampunk vibe.

Probably the biggest modern technical hurdle are visual artifacts—the screen door effect (*SDE*, aka pixilation), the Mura effect or clouding, black smear, chromatic aberrations, backlight bleed, ghosting, FoV distortion, God rays, or circular glare—the list can go on. Each of these minor issues can add up to a substandard experience and loss of presence. In the MR space specifically, better simultaneous localization and mapping (*SLAM*) is another necessary component for a more realistic experience.

In the current market, vendors are making names for themselves by improving certain aspects of the experience while deemphasizing others. The technical improvements that are important depends largely on the use case. Wide FoV and 8K resolution can matter for flight training simulators, while 6DoF and long battery life is important for gamers. The military would value sturdiness and redundancy in HUDs regardless of cost, while a corporation using VR for training may be more cost sensitive.

Like all deep technologies, XR has some hurdles, but the greatest complexities have been resolved. The rapid rise of any new tech evolves from incremental improvements, effectively field-tested at scale. At some point, the details all add up to create something at a quality and price necessary for mainstream success. In the early days of this decade, we're right at the tipping point.

CLOSER WITH XR

Author Kurt Vonnegut once said, "We are what we pretend to be, so we must be careful about what we pretend to be." This quote could easily apply to extended reality. When we're each subsumed into our own custom realities, it raises philosophical questions of intersubjective, shared narratives, and psychological questions of individual health and the social fabric. It even raises religious questions of how we connect to the divine. Fundamentally, we need to ask if XR will improve our lives or if it will further atomize us. Can we work together when we no longer share a reality?

My experiences with XR have changed my relationships in a few important ways, and I hope it will change most of us for the better. Here are a couple of examples of how XR has impacted my life. After only three hours of sleep in a seventy-two-hour period, my daughter was finally born. I'll never forget the exuberance, relief, exhaustion, and fear that arose in that moment. I also felt love, of course, for my girls and the world. What I didn't think of was the headset on my face, recording every minute that I would relive over the years. I was present, and Google Glass AR captured it all. XR brought me closer to my own life.

I stood suspended in the vacuum of space, surrounded by stars in every direction, hovering over a black hole, and jumped. The blue stars shifted, moving faster, the universe bubbled around, and then it was above me. I was outside of space and time. Mathematically, we know the physical stress of gravity stretches your body to a long strand of spaghetti. You cannot survive a real black hole jump. But VR made the impossible possible and suddenly awakened in me a new appreciation for the harshest environment. XR brought me closer to nature.

Extended reality is the most direct interface to our greater technology ecosystem than any other previously devised. By the end of the decade, the action of staring at our smartphones will become passé. In light of smart glasses capable of augmenting reality with digital information, mixing physical objects with virtual objects, and overriding the real world with entirely virtual realities, this may become the dominant form human-computer interface in the next century. At the far end of each piece of technology, though, is another human—a heartbeat connected through digital veins. XR tools can finally bridge us together far more directly than opaque websites and video chats. Any of us can experience a sense of instant presence, designed with our own sensitivities necessary for impact. We can be more engaged, more empathetic, and more connected if we're deliberate. XR can bring humanity and the world we share closer together.

FURTHER READING

- *Ready Player One*. Ernest Cline
- *Future Presence: How Virtual Reality Is Changing Human Connection, Intimacy, and the Limits of Ordinary Life*. Peter Rubin
- *Virtual Reality*. Samuel Greengard
- *Augmented Human: How Technology Is Shaping the New Reality*. Helen Papagiannis
- *The History of the Future: Oculus, Facebook, and the Revolution That Swept Virtual Reality*. Blake J. Harri

4

BLOCKCHAIN, CRYPTO, AND HYSTERIA

It was late fall 2017 at the height of cryptocurrency mania. Bitcoin was a seemingly unstoppable rocket to the moon, eventually hitting a peak of $18,000 US per coin. People sold all of their possessions, including family homes, in a frenzy to buy a few bits associated with an address on a global ledger.

I had flown into Las Vegas for a conference when I first realized this boiler was primed to explode. The biggest online wallet for storing cryptocurrencies, Coinbase, was on pace to become the first legitimate crypto unicorn, based solely on buying and selling an unproven digital currency that was both highly insecure (the successful hacks were too numerous to track) and a terrible currency (a volatile currency that was hardly useful). Upon learning I was in the technology business, my Uber driver from the airport asked if he should buy Bitcoin or Litecoin. By November 2017, those coins were pedestrian choices. When I arrived at my destination, although the conference was ostensibly about cloud technologies, the only topic of conversation was so-called altcoins, namely Ripple (XRP), which was trading at 25¢ per coin. Five weeks later, it traded at $3.80 per coin, earning more than 1,500 percent ROI in the time it takes to order a wedding dress.

It was clearly a bubble, but how large would it grow before it popped, and how far would it crash when it did? For those of us involved in trading crypto since the early 2010s, peaks and long winter valleys were the norm, but it always rallied stronger in the next cycle. We were an insular group, namely trading among ourselves. The most famous early transfer of crypto funds was 10,000 Bitcoin (BTC) for two pizzas, valuing a single BTC under half a cent. Eight years later (before the crash), it would be a $100 million pizza.

By December 2017, cryptocurrency became a staple, from CNBC to Fox Business. Institutional investors began making claims that crypto should be a part of healthy portfolios. Crypto whales (high value individuals) were regularly featured in the *Wall Street Journal*, and the leaders in the field were featured at global economic forums. Times, they were a changin'. At least, for a few months. There was a depression era joke that said if your shoeshine boy started talking about a stock, it was time to sell, and my Uber driver's interest was a clear indicator that the burst was nigh.

When the bubble burst in early 2018, many were financially ruined and dreams were dashed. Benoît Cœuré, executive board member from the European Central Bank (ECB), called Bitcoin "a combination of a bubble, a Ponzi scheme, and an environmental disaster." The party appeared to be over before the headlining act even took the stage. So much for revolutionizing the financial industry and supplanting fiat currencies.

A cottage industry had grown around supporting crypto, but

what to do now? A glimmer of hope remained for those dreaming of selling pickaxes (a common gold rush metaphor for those who got rich, not by mining gold but by selling pickaxes to all those eager miners dreaming of avarice). If the future of cryptocurrency was unclear, what about the underlying technology itself, blockchain? The infrastructure was there, along with the knowledge and expertise. To borrow a phrase from the startup world, the crypto players were "making a pivot" from currencies to blockchain.

To understand the current and future state of the blockchain world, let's first discuss what blockchain is and what it might be used for. Then we'll dig into some technical details of how blockchain provides its unique value proposition, followed by an audit of some of the work left to do. At the end, we'll revisit the question of whether blockchain is the technology we should be investigating or if we should be looking elsewhere for answers.

BLOCKCHAIN, AKA DISTRIBUTED LEDGER TECHNOLOGY

Imagine a huge spreadsheet or ledger of transactions—financial or otherwise. This is a list of every transfer of funds or data from one party to another for all time. If I send a payment from my account (delineated by a unique *address*) to a restaurant's account to cover the cost of a cheesesteak sandwich, that transaction is permanently encoded in this global ledger. *Blockchain* is the technology that manages this ledger, and the currency built to operate on it is a *cryptocurrency*. Blockchain technology was initially invented for the creation and transfer of a nonfiat currency called Bitcoin, which spawned an entire ecosystem of thousands of competing cryptocurrencies of varying values.

What makes blockchain interesting for currencies is that it

creates a stable system without the need of a bank or central authority, where double-spending of funds is impossible. Over time, this technology generalized beyond currency exchange, and we started calling it *distributed ledger technology* (DLT). DLT technology can be thought of as a giant write-only database where data is spread across every computer in the network rather than stored in a central location like a traditional database. This design has a peculiar tradeoff which is the source of its complexity and strength.

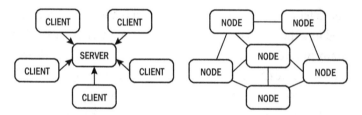

A standard database compared to DLTs—the latter has no single point of failure.

DLTs enforce the fundamental rule of *irreversibility*, where all new records added are *immutable*, unlike a regular database where you can fix and update data. This means that if I write data to the blockchain, it's permanent. The only way to fix a mistake is for the affected party to intentionally reverse it.

Let's say I place an order on Amazon for a pair of shoes and a shirt, then change my mind and decide I don't want the shirt. The standard database that Amazon uses lets me update my order and remove the shirt from my cart, leaving only the shoes. DLT works differently. If Amazon decided to track all purchases on a blockchain, then I'd have to live with my mistake. If I don't want the shirt, I need to create a second transaction later and ship the shirt back. While this a horrible attribute for online shopping carts, it's exactly the kind of behavior you want for financial transactions and many other kinds of data.

A TRUSTED THIRD PARTY

How many institutions exist for the purpose of trust? When I write a check to my dog groomer, they must trust I have the cash to cover the amount, but we are not the only players in the equation. I'm leveraging trust in my bank to pay the recipient from my account. If there's any question about trusting the bank, you can try a direct approach with your next purchase and write your grocery store a personal IOU for some apples. They're unlikely to be thrilled. The trusted third-party bank ensures my debt will be paid. Central banks are a lender of last resort, generally playing a similar role between banking institutions. Fiat currencies are based on trust in the backing government to guarantee the underlying value.

To give other examples, when you buy a house, an escrow agent is a trusted third party for holding funds that are intended to transfer between you and the seller until all terms are met. A notary acts as a third party to verify that you signed your closing documents. And when you visit your loan officer's website, security is guaranteed by a trusted third party called a *certificate authority*, who vouches for the encryption of the data in a page you visited. In each of these cases, blockchain could be the trusted third party, acting as a middleman between you and who (or what) you're dealing with. Rather than trusting a person, agency, or institution, you can instead trust a shared distributed technology network. It's with this understanding that we can investigate some of the use cases where distributed ledger technologies like blockchain could fit, from enterprise industries to consumer products.

SUSTAINABILITY

Sustainability is a hot topic in DLT circles these days because there are so many players in a sustainable system who must trust each other while also competing. Sustainability requires tractability and transparency of goods in a way that can be trusted by industry, auditors, and the public. When the provenance of leather is tracked in every pair of sneakers, it's pretty easy to prove whether or not a particular pair was farmed from a location that contributes to rain forest depletion. If every step in the manufacture of a high-end handbag is tracked from the factory, it's possible to track if the workers are paid fair wages. IXO is one of many vendors in this space, tracking environmental impact of personal and institutional decisions, such as carbon emissions and credits. We'll discuss how it can be possible to ensure this sort of physical-to-digital tracking in the IoT chapter.

SUPPLY CHAIN

In the summer of 2018, the technology company IBM and global logistics provider Maersk announced a deal to cocreate a new supply chain software company called TradeLens. This is normally the kind of announcement that few would bat an eyelash at were it not for the crucial ingredient of IBM's DLT called Hyperledger Fabric. The goal of TradeLens was to digitize the massive amount of paper that exists in the field of international logistics, from bills of lading to customs forms. The true innovation here was not yet another digital system with dreams of global domination (we already had SAP for that) but the network topology of Fabric.

Imagine your job is to convince governments to digitize their legal documents. If you go to Russia and ask them to put their customs forms on a standard computer service, those servers have to be physically located somewhere. Russia is likely to demand that their data lives in servers that reside within their own borders. In fact, it's safe to assume every country would desire a similar option. Since DLT nodes can reside anywhere—unlike a cloud datacenter, for example—this is an easy requirement to meet, and each country would be guaranteed their transaction records are secure, unalterable, and auditable. And finally, the network would need to be safe from nation-state hack attempts without a single point of failure. Luckily, we have such a technology in blockchain. In a very real way, DLTs are one of the best technologies available to help governments trust in international data exchange, which is a huge step forward in digitalizing governments' addiction to paper.

HEALTHCARE

My first job out of college, I worked as a programmer at a large healthcare software company in the middle of Nowhere, USA. What immediately struck me was how inefficient the healthcare industry was and how they were hesitant to integrate health records between individual hospitals or networks. While there were plenty of integration standards (HL7 was all the rage twenty years ago), there was no national database of healthcare records in the US. It's possible there never will be unless patients start demanding it. With the growth of healthcare options, it's increasingly important for individual patients to start keeping their own records, which again, requires a third party to trust. Or alternatively, it requires a DLT to store trusted health records, where the patient can choose which records to share and with whom, independent of any particular health provider or even government. Medicalchain is an organization attempting such a feat. An interesting twist is that it also allows users to monetize their private health records for research purposes. Now *that's* owning your own data.

In the US, 34 percent of healthcare costs are administrative. Anyone who has ever filled a prescription or scheduled a medical procedure knows there is a waiting period between the decision to proceed and the actual medical act. Much of this wait is due to insurance administration, from ensuring the right data exists for a prior authorization to correctly coding the procedure to be covered. A shared and common trusted system for instantly transmitting this data could reasonably decrease healthcare costs by double digits. Moreover, transparency between wildly different costs for identical procedures taking place in different hospitals might allow patients to force price competition and cost alignment into an opaque system.

IDENTITY AND ACCESS MANAGEMENT

Like most Americans, I have a social security card—tattered, stained, and unlaminated. Because the USA, like many countries, lacks a national ID system, this little shredded paper contains an insecure, incremented (and when I die, recycled) SSID. This represents a standard identifier for a multitude of purposes, from attaining a credit card to getting a new job. That SSID is associated with thousands of databases, each one containing different and potentially conflicting data about my identity, from credit scores to criminal history. A central, secure, distributed ledger of major life events that I could track, verify, and share with others only when I provide digital permission would be a major improvement from a smudged and tenebrous paper document. This kind of digital identity is a human authenticity guarantee to others that I am who I say I am. Oftentimes, that proof provides certain benefits (a social security check) or access (entry to a private club).

Beyond people, there are objects that also need to be proven authentic, and rights are bestowed upon individuals or systems that meet certain criteria. Digital rights management via blockchain is a burgeoning area of research. Consider a star athlete or musician who wants to be paid a nominal fee whenever their image is seen or their song is played. Digital ledger technologies can track when a digital file is used and charge for the right to access it. The contract is built entirely around usage. Unless you pay, it's impossible to decrypt or access the file. Moreover, you can pay in real time for partial use. Say you listen to one minute of a three-minute song. A smart contract can extract micro payments against the file stream, and you'd only pay one-third of the total price.

SHARING ECONOMY

Despite not owning a single hotel, Airbnb is one of the most valuable hotel chains in the world. A similar comparison holds for other brokers like Uber and Lyft. One of the more exciting use cases of blockchain is the ability to rent out equipment you own to others, without a middleman taking a cut. Arcade City, a ride sharing app backed by blockchain, is an open competitor to Uber, where anyone can post their intention to drive people around and users pay via token directly on the platform. There's no middleman other than the blockchain. OpenBazaar has slightly broader ambitions, ostensibly taking on Amazon, eBay, or Etsy as an open marketplace where anyone can post or purchase goods. While Amazon takes a 30 percent cut or charges a fee to open a store, OpenBazaar is a completely open platform. The only fees are the small amount required to run the code necessary to post or purchase goods and services.

Use cases for trust abound, and we could fill an entire book on potential opportunities to leverage DLTs. Consider, for example, a distributed ledger of votes that's irreversible, trackable, and anonymous (you get your own secret ID per election) but publicly readable (you can count the votes yourself, if you like). Or a DLT-based system that works for secondary markets for gently used items, such as cars. I could go on.

Thus far, you've had to trust me that this technology can be trusted. Now it's time to dive deeper into exactly why and how we can be sure this trust has been earned by drilling into the DLT that started it all: the Bitcoin blockchain and its *Proof of Work* mechanism.

A GENTLE BREAKDOWN OF PROOF OF WORK

There are many technologies one can understand with scarce understanding of the technical details. Blockchain is not one of them. Its very soul is technical, and understanding its value is an exercise in subtlety. Even if the implementation is complex, the concept is simple.

Fundamentally, Proof of Work (PoW) makes it far too expensive to alter history in terms of physical computing power and electricity. Proof of Work runs on a cross-section of cryptography, statistics, and the physical reality of what humans can reasonably compute.

When a certain number of transactions (such as sending two Bitcoins to my mom's account) are made on the network, computer nodes called *miners* attempt complex math puzzles to bundle up the most recent transactions into a new *block*. If you think of transactions as rows on a spreadsheet, a block is basically a certain number of rows in a single file. The miners race to collect transactions and "sign" the next spreadsheet. The winning computer in this contest gets to permanently lock out all other nodes from making changes. So, when a miner wins the contest, the remaining nodes move on to create the next spreadsheet in the chain. The sequence of all blocks creates an ordered series of transactions across the planet called the *blockchain*. It is, quite literally, a digital chain of blocks.

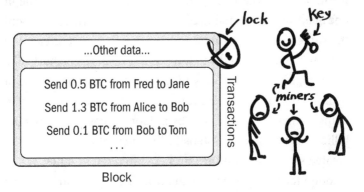

Miners compete to be the first to discover the correct, completely random key.

WINNING THE BLOCKCHAIN LOTTERY

When a miner collects a list of transactions to build a block, they encode all of this and other data as a *hash*. A hash is just a big number, common in cryptography, that can take any amount of data and encode a single fixed value. For example, the word *breakdance* hashes in the SHA256 algorithm to: B0A9E90D61F43A0F5166593523CD6B661AA27AE7D871AD B3A1CF4B4C67E3C5CA.

If you were to hand this number to someone, they could never convert it back to the word *breakdance*. It's a one-way check. However, if you were given the word *breakdance*, you could always generate that long hash number exactly the same way every time. When slight changes are made to the block (via an incremented digit called a *nonce*) and the correct hash is generated according to certain rules (for example, the number must start with eighteen zeros), the miner wins.

Let's consider a game where you throw fifty pennies into the air, and they all have to land on heads to win. The odds of any throw working out in your favor are 1 in 2^{50}, or about one in one quadrillion (one followed by fifteen zeros). These are about the same odds of a miner winning the chance to create the next Bitcoin block on the first try. If it sounds suspiciously like a lottery, it's because it very much is. Although the chance of winning the race to make the next block has low probability in a single attempt, multiply this by millions of computers making billions of attempts, and statistically, a new block is likely to be created somewhere on the internet within ten minutes. When a computer finally generates the winning hash, it informs the rest of the network, where it's easy for other computers to verify the winner.

The integrity of the blockchain is proven due to the sheer amount of *work* necessary to generate a new valid block that hashes correctly—hence, *Proof of Work*. It takes a tremendous amount of computing power, and thus electricity, to stumble upon the right block hash. That's the brilliance of the system: it's far more complex to generate the winning number than it is to validate that the number is correct, so faking history is nigh-impossible. But verifying history is trivial.

FROM BLOCK TO BLOCKCHAIN

Now that we've created a block, how does that make a chain? Blocks are not exclusively a spreadsheet of transactions. They also contain other data about the block itself and its place in the network, called the *block header*. This metadata is just as important as the transaction ledger data.

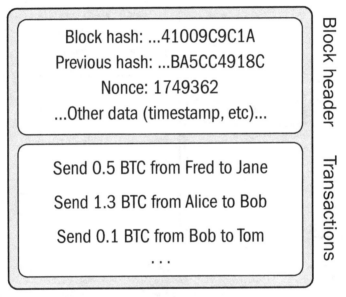

A Blockchain block contains a list of a few transactions, organized by a header.

The block header contains mining values (like a time stamp, number of transactions, and a nonce), a roll-up value of the transactions, and most importantly, it contains the *hash of the previous block*. The current block points at the hash of the last block, which itself points at the hash of the prior block, and so on. It's like a backward conga line, where everyone places their hands on the hips of the person behind them rather than on the person in front. Every block points at the previous block, making a chain all the way back to the very first block ever created, called the *genesis block*.

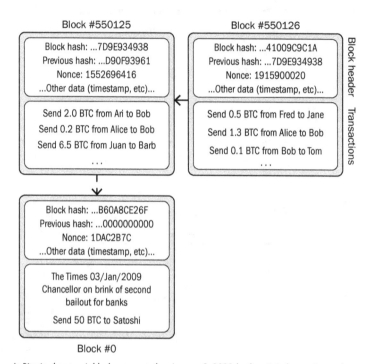

Block #550125

Block hash: ...7D9E934938
Previous hash: ...D90F93961
Nonce: 1552696416
...Other data (timestamp, etc)...

Send 2.0 BTC from Ari to Bob
Send 0.2 BTC from Alice to Bob
Send 6.5 BTC from Juan to Barb
. . .

Block #550126

Block hash: ...41009C9C1A
Previous hash: ...7D9E934938
Nonce: 1915900020
...Other data (timestamp, etc)...

Send 0.5 BTC from Fred to Jane
Send 1.3 BTC from Alice to Bob
Send 0.1 BTC from Bob to Tom
. . .

Block header | Transactions

Block hash: ...B60A8CE26F
Previous hash: ...0000000000
Nonce: 1DAC2B7C
...Other data (timestamp, etc)...

The Times 03/Jan/2009
Chancellor on brink of second
bailout for banks

Send 50 BTC to Satoshi

Block #0

In Bitcoin, the genesis block was created on January 3, 2009, by the original mysterious author, who wrote under the pseudonym Satoshi Nakamura. As proof that the block was created on or after January 3, 2009, it contains the following note: "The Times 03/Jan/2009 Chancellor on brink of second bailout for banks." For a currency meant to disrupt the financial status quo, it was quite a comment on the fractional reserve banking structure.

So that's it! You've now built a chain of signed blocks that are reasonably impervious to attack by humans on Earth with modern technology.

To recap thus far, in severe jargon: *miners* gather *transactions* into a *block* and iteratively increment a *nonce* in an attempt to generate a winning *cryptographic hash* while ensuring the newest *block header* contains the hash of the *previous block*.

This linked list of blocks creates a *write-only ordered ledger* of all *transaction* records, representing the transfer of *cryptocurrency* from one *address* to another. The entire list of ordered blocks starting from the *genesis block* is called the *blockchain*, which is a kind of distributed ledger technology, or DLT. Don't worry if you blanked on some specific terms; it's the concept that matters.

WHO ARE THESE MINERS?

We still haven't explained why miners take part in this work. The short answer is money, because the reward for mining a block is payment in Bitcoin. Mining is not only how the transaction ledger maintains integrity, it's also how new currency is introduced to the network. When a miner wins the block creation lottery, the miner is rewarded by having new bitcoins created and "deposited" to its own address. The reward for generating a single winning block can be worth tens of thousands of dollars, with a new lottery happening every ten minutes. This is a billion-dollar industry.

Let's pause to consider the brilliance of this system. Because of the difficulty in mining a new block, it's effectively impossible for anyone to modify records encoded in the network. Since each block contains the hash of the previous block, the only way to rewrite history is to rebuild every block for all time, which isn't physically possible on the Bitcoin blockchain, given the world's computing power by cost. Furthermore, because there's a potential financial reward for taking part in verifying transactions, there's a global network of computers fighting for the chance to protect all transactions. Because those miners are rewarded in Bitcoin, they have a financial incentive to ensure the integrity of the system as a whole.

If any particular group became greedy and decided to mine so much that they effectively owned a majority of the network, it could destabilize the value of Bitcoin since users would no longer trust it. The miners would then have a majority share of a worthless asset. In other words, everyone has an interest in taking part in the network and protecting its integrity. This is a network designed not merely on cryptography (mining) and statistical likelihoods (block creation will eventually happen somewhere), but it is also rooted in game theory and, ultimately, human psychology. The more stake an individual has in the network, the higher their incentive to ensure its integrity.

With this basic structure in place, some folks realized that the Blockchain ecosystem could be leveraged for more than storing data through predefined transactions. It could be further generalized into a global network of general purpose computing resources. This is the turning point in our story where everything changed.

DAPPS

The blockchain use cases for most early cryptocurrencies are similar and emerge from a sort of libertarian appeal to avoid central control and fiat manipulation at all costs. The blockchain is public, ensuring it provides a radical transparency into all transaction history and ensuring no alteration has occurred. The network must also be trustless, which means that all computers on the network are equal, and anyone can take part in trading or operating (mining for) the network. There can be no *Animal Farm* scenario, where some computers are superior to others. No miner is privileged above any other; there is no central authority or control.

But humans are still involved, so philosophical conflicts occur. These disagreements between experts are often resolved by making changes or "forks" in the code that miners choose to run, effectively creating new currencies. This has happened many times. Most famously, a disagreement between groups of programmers in the Bitcoin protocol forked the blockchain into two competing currencies: Bitcoin (BTC) and Bitcoin Cash (BCH). But even when a fork in the codebase happens, there's no loss or rewrite of history; it's just two roads diverged in a wood, and different people choose different paths going forward.

It's against this backdrop in 2013 that a nineteen-year-old programmer and Bitcoin enthusiast named Vitalik Buterin had an idea. Rather than trying to make the core protocol of Bitcoin everything to everyone, why not create a blockchain network that was capable of running custom software? Furthermore, why not let the miners earn money executing useful code instead of merely racing to see who can generate the biggest random number the fastest? Sure, there would still be mining, currency transfers, and block creation, but there would also be a readily available computation platform where anyone could pay to execute *distributed applications* (aka dApps). Unlike the Bitcoin blockchain network, which only traded Bitcoins, anyone could code their own tokens, bound by whatever rules they wanted.

In 2015, the network launched by Buterin and others was named Ethereum (née Frontier) on the back of a crowdfunding campaign that raised $18 million USD. Now this network has its own cryptocurrency named Ether (ETH) which can be traded, similar to Bitcoin, but can also be used as a reward to miners who were willing to execute code. Think of the Ethereum network as a giant global virtual computer made up of millions of physical computers, ready to execute any code you pay for them to run. Because of this view, the Ethereum execution network is called the Ethereum Virtual Machine, or EVM. Anyone who knows how to program an EVM scripting language, Solidity, can write code that others can run on the network. To ensure idempotency, the source code is stored on the blockchain, just like a financial transaction might be.

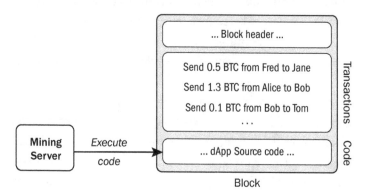

Code is stored as data in blocks, and miners can bid to execute it for money.

These dApps are public, and execution is irreversible once someone has chosen to run the code. For those with a broad imagination, they are akin to self-executed legal contracts, earning them the name *smart contracts*. Since smart contracts are just computer code, there are few limits on what behaviors you can code into them. You can set up a smart contract where any user can rent a car, and while they drive the car, the contract will continuously transfer funds to the car's owner until they remove the key. You can create a smart contract that attaches to a beer tap and charges per pour. You can create a smart contract that generates images of cartoon cats and sells those images on a marketplace of cartoon cat breeders. Seriously, this is a real thing, called CryptoKitties.

THE DAO

In early summer of 2016, some of the core creators of Ethereum had a prescient idea: if you could codify a sophisticated enough smart contract, could you design an organization that is largely run autonomously? Any interface between the organization and individuals would be dictated by this contract, effectively rendering the organization as little more than a set of rules, a pool of currency, and a governing body whose governors were elected by shareholders, limited to actions explicitly stated in the contract. Anyone could be a shareholder merely by putting money into the system, through execution of a public smart contract. Any payment of funds for work would require approval by the governors, and such officers could be removed and replaced at any time by a shareholder vote. The organization was a *distributed autonomous organization* (DAO).

This large public offering had some analogs to the kind of initial public offering (IPO) of stock that you would expect to buy through a stock exchange like NASDAQ. Since your ownership in the company took the form of tokens or coins in the Ethereum network rather than stock certificates, this kind of public crowdfunding of a blockchain-based business became known as an *initial coin offering* (ICO).

The DAO ICO was an exciting time and heralded both a new funding model and a new form of business. In a world where Worldcom and Enron executives became wealthy at the expense of shareholders, such total transparency was a breathtaking experiment. The DAO raised 11.5 million ETH from over 11,000 individual investors (including yours truly), valued at $150 million US at the time.

What the DAO really was, ultimately, was an exercise in perfect bureaucracy. No one was above the law, but there was also no real authority to appeal to when something went wrong. In a regular contract, honest disputes can be resolved by courts, but what recourse is there against a conceptual flaw in a contract that is publicly executable and irreversible? The fundamental error of the DAO centered around a bug—not in technology or software—but in logic. A user found a flaw in the DAO contract and drained the account of nearly $50 million US before the community was able stop the bleeding. Many loathe to call this a hack or a theft because this person's actions were not so clear cut. He or she did not break any rules of the system as designed but merely outsmarted the thousands of people who had put funds into the DAO account. What the user did was entirely within the bounds of the contract as designed.

Following a rapid community debate and vote on what to do next, the core Ethereum developers opted for the nuclear option: to fork the blockchain and return funds to the investors. This fork created two currencies, becoming the mainline Ethereum (ETH) for the majority, then Ethereum Classic (ETC) for those who believed that maintaining the integrity of the blockchain was more important than $50 million. Ironically, a blockchain designed to avoid forks was ultimately forked.

THE RISE OF THE ICO PHOENIX

The failure of the DAO should have been a black eye for any emerging technology, but history has shown that this loss of funds deterred no one. The takeaway for the community wasn't one of tempering irrational exuberance (eighteen months later, ETH was trading at a 13,000 percent increase) but rather that large sums of money could be raised without any product at all. If just an idea and a smart contract raised $150 million in an unclear experiment, imagine how much could be raised with a bigger idea! And thus, the ICO boom was born—a phoenix from the ashes of the DAO, fueling a two-year race to raise ever larger amounts of funds with very little effort.

In 2018, ICO mania drove an overall market of untested startups to raise over $13 billion US. The most sophisticated single project, named EOS, raised over $4 billion US to build an alternative to the Ethereum network. The jury is still out on whether or not they'll be successful, but it was an attractive option for those who felt they had missed out on the early days of ETH and BTC. (FOMO, the fear of missing out, was a powerful driver of many ICOs). Initial coin offerings blurred the line between holding stake in a company (a security), holding tokens that could be transferred for goods and services, and holding an asset like Bitcoin that could be used as a general currency.

The DAO launched the mania for companies to run on smart contracts, and people wondered why they should go through all of the work of building a company when they could just launch an alternative currency and hype it on Ethereum? Thus, a new set of coins emerged, collectively called *altcoins*. The sets of coins that ran on the EVM were so similar, the ICO community even created a standard protocol called ERC20. This ICO mania was inexorably linked with the hunger to invest in more cryptocurrencies, which fueled the creation of more altcoins. For a brief time in 2017, anyone with a little programming ability and the gift of hype could be a millionaire by merely copying an existing altcoin smart contract and creating a slick website to sell the new currency. Over the span of a few months, most ICOs failed, either due to incompetence or outright fraud. This placed a chilling effect on market exuberance and on the value of blithely generating new coins.

With the promise of instant millions becoming a fading dream, and the value of cryptocurrencies plummeting, was the market dead? Was there nothing left of this grand experiment? It wasn't quite time to despair because, as we've already explored, there are other uses for a write-only distributed ledger technology. However, there are some technical problems we need to work through before DLT is ready for a broader adoption.

AN IMPERFECT STORM (CURING THE ILLS OF DLTS)

We ran through a few use cases for DLTs, and there are countless more. We checked out how a type of DLT works and how that expands into dApps. We also glossed over the many weaknesses of the DLT technology and how the greater blockchain industry is working to resolve them with varying degrees of success.

TRUSTING THE PHYSICAL WORLD

The power of blockchain relies on the fact that any record placed in the blockchain cannot be altered, mutated, or reversed. But we have to keep in mind the old computer information mantra: "garbage in, garbage out" (*GIGO*). Just because blockchain ensures data is irreversible, how do you trust that any data put into the system can be trusted? If someone creates a blockchain that tracks my heart rate in order to get a discount on life insurance, it's only useful if my vitals are read by a device that can be trusted to be correct; its results cannot be tampered with, and it can securely submit those results to the blockchain. If the underlying blockchain requires me to manually submit my vitals on a monthly basis, I have every incentive to lie if I have bad results so I can get the discount. It has to be secure all the way down.

When blockchains are populated by external values, the trusted systems that convert data from the real world to the blockchain are called *oracles*. Oracles are only useful insomuch as they can be trusted, so they tend to enforce that trust with encryption that can only be decrypted by the smart contract or by secure hardware that ensures the values haven't been tampered with in some other way. Anything less than total round-trip security can open up the blockchain to garbage data, which makes it difficult to rely on. GIGO. One of the more recent companies looking to solve this problem through secure hardware is Chainlink. It's an interesting option for tying IoT data to a blockchain.

THE PROBLEM WITH LARGE FILES

There's a lot of talk about blockchain for trading large files, from healthcare records to digital contracts. But think back to how blockchain actually works. It's much more akin to a spreadsheet of transactions than a large file storage like iCloud or Dropbox. If someone creates an application that stores legal documents somewhere on a standard cloud service, and signing that document creates a record in blockchain, what stops me from modifying that document directly on the Cloud Filestore or just outright deleting the file? Sure, the signature is there, but what exactly did I sign? Just like our problem with trusting the physical world, we need to trust that large files will be distributed, always accessible, and that changes are irreversible.

There are a few blockchain-based solutions for this issue. The first was *IPFS*, or Interplanetary File System. What makes IPFS interesting is that it starts with a blockchain-like write-only data structure (Merkle DAG). Merkle DAGs leverage the best of two worlds. They leverage the insurance of DLT's so all changes are kept forever, and they also leverage technology beyond BitTorrent, allowing a file to be broken into bits. Put these together, and the community of storage nodes can reconstruct any stored file. Servers store pieces of various files and receive the most reward for holding onto the rarest blocks, ensuring that every piece of every file in the network is held somewhere by some computer. Reconstructing those pieces is as simple as asking the network where those blocks are and connecting them. Storj is another large file system, along with a handful of others, and what matters most is that these file stores have features similar to blockchain: distributed and irreversible. As long as such large file stores exist, blockchain can point to the file and version it's transacted against.

BLOCKCHAINS ARE SLOW

One of the oldest concerns about the Bitcoin blockchain is how slow it is to resolve a transaction. The Bitcoin network famously supported nine transactions per second, and each of those transactions took ten minutes or more to resolve. Many fixes have occurred over the years, with various solutions. There are sidechains, which are basically standalone blockchains with custom rules that can eventually converge into the main blockchain. There is the Lightning network, a type of sidechain dedicated to supporting many rapid transactions between a limited number of traders, whose final result merges into the main chain. The Ripple network can validate a transaction in four seconds. Zilliqa has a blockchain that can validate 15,000 transactions a second. The credit card company Visa can support 24,000. An emerging blockchain called Ternio is developing an implementation that can cover one million transactions per second. Other laboratory tests have even higher rates, one even claiming seven million transactions per second. All of this indicates that the famous slowness of early implementations is being resolved, while blockchains are still distributed, irreversible ledgers.

ENVIRONMENTAL IMPACT

This one is a big deal. While the benefits of a blockchain architecture are myriad for some of the use cases we've outlined and many more, it's hard to make claims that DLTs based on Proof of Work are good for the environment. As of 2018, executing a single transaction took the same amount of energy and power as about twenty-five households for a day in the US. Based on the given rate of Bitcoin transactions, that is equivalent to the energy consumption of the entire country of Austria. Essentially, the lottery system we previously discussed makes it difficult for someone to generate a block, theoretically giving everyone a fair shake at making the next one, based statistically on how much electricity they're willing to burn. But there are other ways of creating new blocks, generally called *consensus protocols*.

The top alternative contender is called *Proof of Stake* (PoS), which rewards those who are willing to put the most stake (aka cash) into the election system. The theory is those who have the most to lose are most likely to ensure the integrity of the system. If you act badly, you'll be punished financially. What's considered "bad" depends on the system, but commonly, it's bad to vote against the common consensus. In Casper, the PoS system for Ethereum, the set of conditions that punish bad actors is called *Slasher*. Then there is *Proof of Authority* (PoA), which gives a sort of political bend to PoS, where stake and being a good actor earns you blessed authority, giving you a right to vote.

Similar to PoS, acting badly means punishment via loss of

stake as well as being demoted from blessed authority. Another method of leader election championed by Intel is called *Proof of Elapsed Time (PoET)*, where candidates attempt to solve secure puzzles in a given time frame to prove their worthiness. Solving those puzzles requires specialized trusted execution environments. Intel happens to manufacture these in a chip form called SGX. There are other proofs, like Proof of Activity, Proof of Burn, Proof of Replication, and so on. While it's still too early to tell which, if any, will emerge the victor, there is plenty of work being done in the space to cure blockchain of its addiction to electricity.

PUBLIC AND TRUSTLESS

While public tractability of all transactions is a good feature for promoting trust in cryptocurrencies, it's not the best quality for a DLT system designed to store personal healthcare records. And while the trustless nature of blockchain is the killer feature for Bitcoin, an intergovernmental platform shouldn't allow just anyone to join their customs data network. The first decade of DLT was all about adding features to a *public, trustless* blockchain. Now academic and industrial eyes are focused on building *private, trusted* blockchains. While the technical details of how they differ can vary, the use cases tend to be more aligned with the need for large organizations (like nations and industrial players) to share data with each other in closed marketplaces. *Corda* and *Fabric* are two leaders in this space. Many private blockchains look to leverage the mindshare of the existing developer community by making available the same smart contract interface of Ethereum's Solidity code, but repurposed for a closed infrastructure.

While it's probably correct to think these blockchain implementations are antagonistic to Satoshi's promise of radical transparency and decentralization via blockchain, it's also possible that these kinds of quasi-private consortiums will have the biggest and longest impact long term. While the jury is still out on the world's hunger for a hard digital currency with no government backing, the future is bright with possibilities for DLTs supporting increased transparency between large organizations and states.

THE END OF VERTICAL INTEGRATION

In the early days of 2019, cryptocurrency values were dropping to an unknowable floor. The majority of ICOs are failures, and fraud in the blockchain space looms rampant. This is in addition to the environmental impact of Proof of Work and the remaining technical unsolved issues that plagued DLTs. Moreover, the space is still so nascent and misunderstood that it's hard to grasp specific use cases that DLTs could better solve than existing distributed databases. You'd be forgiven for harboring a sense of futility around the whole exercise.

It's important to separate the turgid, get-rich-quick hype of cryptos fueled by ICOs from the technical guarantees of DLTs. There are clear use cases for distributed irreversible ledgers ranging from personal conveniences to governmental reform. One exciting change is how DLTs can evolve enterprise companies of the future into smaller, leaner, distributed networks within well-defined contracts as code.

The nearest term value of blockchain rests in leveraging the technology to build consortiums of companies with a common alignment. There's relatively little risk in joining, for example, a product authenticity tracking group to fight black-market counterfeits, or taking part in a sustainable circular economy for cotton production. This first wave of consortiums were a first pass at decentralization, allowing companies to relax centralized control with low risk. The benefits of giving up certain rights of governance to access the economies of scale of an orderly cooperative were too good to pass up.

As organizations become more comfortable joining networks of other organizations and are backed by a technology infrastructure that makes such consolidations effortless, this may expand into increasing core operations that were once vertically integrated in favor of open marketplaces. For the buyers, exposing the details of their operations will allow providers to better service the enterprise, while membership for the sellers evolves from competitive advantage to necessity.

Consider the case of transportation. Why would a grocer choose to have a contract with a single large-scale shipping company when, instead, they can place bids on a marketplace populated by a consortium of transportation agents at a lower cost, higher quality, and with full transparency? Each agent has a set of services they can provide and a history of outcomes; the buyer merely places a transportation bid against the consortium's smart contract with certain defined requirements and accepts the lowest qualified bid. With such an open marketplace, why would a driver choose to stay with Uber when they could work directly with the buyer? Uber would eventually need to join this consortium or at least create one of their own, perhaps with other major players. A company can't push back forever against the relentless force of strong atomized networks.

While the DAO (distributed autonomous organization) experiment didn't immediately work for those of us who invested early, the concept can't be dismissed out of hand. Organizations will be increasingly distributed, and such distribution will be held together by autonomous operations. Today, robot process automation (RPA) is happening globally and giving way to an AI-based version called cognitive automation. As more operational processes are internally automated in increasingly intelligent ways, eventually that automation will expand into operational tasks that interface with external agencies. The increasing automation will require a system through which companies can engage with rules that enforce good behavior. There will always be cheats, and AI has been shown to be one of the greatest cheaters in history. So the system itself needs to keep an irreversible ledger of all historical operations and enforce certain rules, for which smart contracts would be a reasonable fit.

Finally, we have to admit, it's possible that blockchain is not the right fit as a technology for many of the use cases we've discussed. However, we do and will increasingly require systems that can act as trusted third parties by allowing data to be shared between individuals and organizations that may not necessarily have honest intentions. In this case, the system itself must enforce its own rules, and such a system requires the trust of its users. That trust can be a collective that everyone agrees has the right to act on everyone's behalf (i.e., a governing body) or a technology system that, by its very nature, disallows all but very specific behaviors (i.e., smart contracts, dApps). While it's very possible that what we currently call blockchain may be antiquated in a few years, the need for some kind of smart distributed ledger technology is likely to persist, so it's worth pursuing now. Since the future of corporations will trend toward more networks and relationships between entities, at the very least, blockchain can act as the training wheels for this inevitable shift.

FURTHER READING

- *Blockchain Revolution: How the Technology Behind Bitcoin and Other Cryptocurrencies Is Changing the World.* Don Tapscott and Alex Tapscott
- *Smart Contracts: The Essential Guide to Using Blockchain Smart Contracts for Cryptocurrency Exchange.* Jeff Reed
- *Bitcoin Billionaires: A True Story of Genius, Betrayal, and Redemption.* Ben Mezrich

5

INTERNET OF ALL THE THINGS

"I think there is a world market for maybe five computers."
—THOMAS WATSON, PRESIDENT OF IBM, 1943

(Some quotes age like fine milk.)

Folks in my line of work love to make predictions. Given that, many wouldn't have guessed even five years ago the scale and speed of the IoT revolution. It's taken many of us by surprise. The dream of ever-present automation has been tried and failed countless times since the industrial revolution. We can muse at the sixties space-age cartoon *The Jetsons*, where video-communicator watches and 3D printed meals seemed so specialized and bordered on the absurd. Now, we live in a world where computerized devices like smart cutting boards and water bottles—while not the norm—are not science fiction. The Internet of Things is here now, working in our daily lives in a myriad of invisible ways.

In my house, smart light bulbs slowly illuminate every morning at 5:30 a.m. When I awake, before I even greet my wife, I'll grab the iPhone and check for early morning emails or meetings, then slap on an Apple Watch. I slide my feet into a pair of Nike Adapt BB self-lacing shoes and head out for my morning run. Arion smart insole pressure sensors track how my feet strike, which I study to adjust my gait. I like to wear my AirPod smart earbuds while running, streaming music or audiobooks from the smart watch via the embedded LTE cellphone data network. No phone is required here, and the watch tracks my speed, route, and calories burned. It's also, of course, a watch. When I return home, I step onto my Withings smart scale, which syncs my daily weight fluctuations with other health apps.

Post shower and dress, I clip on an OrCam, an ever-present smart camera that tracks the faces of everyone I meet in the day. It's helpful since I have a hard time remembering names. I head to the kitchen and ask my HomePod smart speaker to continue the playlist I had started on my run. I grab a breakfast bar and scan the barcode on the packaging with the MyFitnessPal app, which not only tracks the calories I intake but also adjusts requirements based on my daily weight and the calories burned on my run. My wife shows me what we were doing at this time a year ago via Timehop, an automatic and personalized "this day in history" tracker. When the last person in my family leaves the house, our Nest smart thermostat turns down the HVAC to save energy, and any stray smart lights in the house will deluminate.

The Waze traffic application raises a collision alert and suggests an alternative, custom route circumnavigating the accident. I swing by Starbucks in route to work and pay with a QR code on my watch that is connected to my Starbucks account, which in turn is connected to my Apple Pay account. As I sit in the drive-through, I call into a Zoom telepresence meeting with Europe. I bid them good evening. The meeting notes are automatically dictated by Voicea, a natural language processor. I pull into my office parking garage where free spaces are illuminated by a faint green light, while occupied spots glow red. I wander toward my office building, whose entrance is controlled via security turnstile. I'm granted access due to a near-field communication (NFC) chip embedded into my employee badge. My phone alerts tell me a package has arrived at my front door, complete with a photograph of the delivery worker, compliments of my Ring smart camera.

I go for lunch at a local fast casual chain restaurant and order a salad on the touchscreen at the table. After the meal, I tap my watch to the face of the tablet, paying at the table with inbuilt NFC and drive home. At home, I pack my bag and throw in a Tile device that lets me track my luggage from anywhere. I ask my AI assistant Siri to order a Lyft to the airport. At the airport, I check my luggage and tap my fingerprints on a biometric and facial scanner called Clear. This lets me skip the first half of the security line. The Clear account is also connected to my lounge account, so again, I press my fingers to the plate outside the lounge to gain access, and I walk inside.

Just as I get comfortable, a tide of terror rises in my stomach, compelling me to check my iPhone. Okay, my front door is locked, according to my August smart door lock app. I press the lock button anyway, just in case. While I'm at it, I use the BMW Connected app to ensure my car doors are locked too. At boarding time, since my digital boarding pass is connected to my smart watch via Wallet, I simply scan my watch to board the plane. I haven't used a paper ticket in years.

Taking my seat, I make myself comfortable and press the attendant button by the seat, and he brings me a water. Is the button IoT, I wonder? I check the status of my Tile, and see that my luggage made it onto the plane as well. Before takeoff, I video chat with my kids on FaceTime and, just for fun, turn on augmented reality mode so my face looks like a puppy for their amusement. We bid adieu, and the plane takes off. I dig my Oculus Quest VR goggles from my carry-on bag and slip them over my head. I'm instantly transported from a sardine can in the sky with thirty-two inches of legroom to a wide open virtual world where I relax and catch virtual fish. For a lifelong claustrophobe like me, this is heaven. I doze off, entranced by the gentle calls of virtual bullfrogs, crickets, and wind rustling in the trees.

The devices illustrated above are not fictional (although the dense usage in a single twelve-hour period was marginally contrived). All of the IoT mentioned are currently used in my household in 2020. We are not particularly gadget-crazed, either, yet the number of devices we use in 2020 are in the dozens, an increase from perhaps a smartphone and some apps a decade earlier. Futurist Kevin Kelly's prediction seems to be coming true. He says that in the future, "ninety percent of your coworkers will be unseen machines. Most of what you do will not be possible without them. And there will be a blurry line between what you do and what they do." The Internet of Things is here, today, and is pervasive.

INTERNET OF WHAT NOW?

Understanding the Internet of Things (IoT) is like learning the game of Go. The rules are simple but grow deeper and more complex as you learn. IoT is (usually) a digital device that connects to the internet (generally), but the Devil is in the parentheticals since many definitions of IoT include technically correct devices like smartphones and technically incorrect passive objects like radio frequency ID (RFID) tags that live inside your passport. And if we include smartphones, why not include tablets, which are basically large smartphones? And if we include tablets, why not laptops or desktops? And if we include RFID, why not QR codes, which also require external sensors to be useful? Where does it end? We'll delve a bit more into the devices and components of IoT later in this chapter (in the "Technology" section), but for now, let's not concern ourselves with an ironclad definition of IoT. Instead, let's be prudent and consider the Internet of Things as a paradigm—we're in the IoT age.

I first came across the concept of the Internet of Things at a global tinkering event called Maker Faire in 2009. *Makers* is the self-applied moniker to electronics-focused hobbyists living in the cross section of open source (freely available tools and documentation) and cheap, available electronics. Makers, as a group, are the very definition of nonexpert early adopters, building intelligence into objects like internet-connected Coke machines or light-up hats for personal utility or artistic expression. The devices that Makers build make dumb things smart. IoT basically achieves the same thing. IoT is, at its core, a movement to extend technology to sense and act in the real world, leveraging network infrastructure (from Wi-Fi to the cloud) to share data.

It would be a stretch to proclaim that Makers are *the* driving force behind IoT; they are a particular intonation of the zeitgeist prevalent in the early twenty-first century. The drivers of Makers and IoT engineers are similar, however: the commoditization and standardization of cheap electronics at scale, the web lowering the bar of expertise for creating new devices from these electronics, and making network infrastructures readily available. Consider the Michigan Micro Mote, a computer that fits on a grain of rice measuring 2 × 2 × 4 mm and powered by a solar cell. In a world like this, almost anything can be made "smart." Installed IoT units jumped from around 5 billion in 2016 to 10 billion in 2018, and to 22 billion in 2020, or to about 2.7 devices per human on Earth. As this growth rate continues, we could easily see well over 500 billion units by 2030, or around 75 IoT devices for every man, woman, and child.

IoT polymath Bruce Sinclair once said that the world does not want a better mousetrap; customers want a pest-free environment. People care about outcomes, not products. Successful IoT leverages technology to achieve better outcomes, while failed attempts try to tie in technology while lacking clear benefits. The value proposition of smart thermostats is clear—you save energy and thus money. It's not so clear for digital bottles that display their liquid contents; that can be also be done by transparent glass bottles. So, in order to understand the IoT world, we'll focus on the big outcomes we want, starting at the micro level and working up to the macro—from personal IoT devices called wearables to smart homes, up to an interconnected smart industry with smart cities, a smart world, and beyond.

WEARABLES, HEALTH

Wearables are a category of the Internet of Things, where you are the "thing" being "interneted." You can wear smart watches on your wrist, smart pendants around your neck, clothing via smart fabrics, smart shoes, glasses, earplugs, as well as clipable devices from pedometers to cameras and microphones. The purpose of many wearable devices is to track steps, heart rate, and blood oxygen levels, collected by enthusiasts who call their measurement methods the *quantitative self*. The biggest market for wearables today is in the consumer health market, but many off-the-shelf wearables are growing in spaces generally dominated by medical devices, such as tracking irregular heart fibrillation or blood oxygen.

The most common general purpose wearable that your average consumer uses is a smart watch, such as the Apple Watch, Samsung Gear, and Fitbit. The watch is a form factor that people are already familiar with, and the bulky size allows space for loads of embedded electronics and a good sized battery. Outside of the health sensor market, smart watches are also extensions of smartphones, complete with smaller versions of apps. During the COVID pandemic, some smart watch apps were used to promote social distancing by buzzing if two or more people came within a six-foot radius. Later, a UCSF TemPredict study partnered with smart ring maker *Oura* to discern if various biomarkers could indicate early prediction of coronavirus, something the device was not originally designed to do.

Amber Case, author and researcher of nonintrusive (Calm) technology, once claimed that we are the first generation of cyborgs. A portmanteau of "cybernetic organisms," cyborgs happen when there is a seamless harmony between living creatures and technology, and the technology occupies a special place in the mind. The technology becomes an innate part of the user, like tools in the hands of an expert. Since a majority of humans have started to carry and interact regularly with smartphones, we are all cyborgs now, and it's only becoming more intimate with newer IoT.

There's an emerging subspecialty of wearables we could call *ingestibles*. Ankon Medical Technologies created a camera pill that can view and measure inside the human body, and a company called *Three Square Market* made headlines by micro-chipping forty employee volunteers to use implants as their ID badges for work and to make purchases. An extreme form of quantitative self-enthusiasts exist in the cross section of Makers and the body modification subculture called *biohackers*: they implant devices such as microchips and sensors under their skin. As bizarre as this may feel to us in the early twenty-first century, by the twenty-second century, this kind of modification may be as mainstream as pierced ears.

From personal GPS trackers to LoJack for your kids or your pets, to a frog baby monitor my wife and I attached to our sleeping infant to warn for SIDS, wearables are an endless subspecialty that could easily fill a book. There are smart bike helmets, sleep and posture monitors, and bionic legs with surgically infused neural interfaces allowing patients to feel sensations of walking, impact, and bending a knee. Many of these wearables are connected through smartphones and share data through common platforms like Apple HealthKit, Google Health, or Nike+.

SMART HOMES

My first unequivocal home IoT device was the Amazon button. Shortly after our first daughter was born, my wife and I found ourselves constantly ordering many diapers and baby wipes. These buttons were connected through our Amazon account and our home Wi-Fi to the internet, placed near the baby changing station. As we ran low on either item, we pushed the button. Within a day, we'd receive a fresh shipment of baby accoutrements. It was a literal easy button.

Connected smart homes are often the first experience people have with IoT devices. Whether it's an Amazon Alexa voice-operated device, a Nest smart thermostat, or Ring video doorbell unit, an increasing number of consumers are opting to simplify some aspect of their lives with smart IoT devices in the home. As of 2020, Americans save well over $100 million per year with smart thermostats. Smart security systems are driving down price and increasing accessibility of top-notch, high-definition security cameras and equipment.

Many of these devices are connected through centralized platforms such as Google Nest, Apple HomeKit, and Amazon Alexa to interoperate in a common ecosystem. These platforms are commonly fronted by voice-controlled chatbots, triggered by branded wake words like "Alexa" or "Hey Google," and followed by a verbal command that is converted to text internally via natural language processing (NLP). Their wide range of functionality accounts for their recent rise in popularity. When I tell my Apple HomePod, "Hey, Siri, goodnight," it turns off the downstairs lamps and TV, turns on my bedroom smart lights at 25 percent luminosity, and locks all external smart locks.

Like wearables, the smart home revolution is expansive. The examples are endless, with more types emerging every day: Eve Aqua home water controller, Athena IoT security cameras, and iBaby baby monitors. In the bathroom, there's the Ayi smart mirror, Kohler's Moxie shower head with speakers, and Withings smart scale that tracks daily weight fluctuations and calculates BMI. In the kitchen, there's the Samsung smart fridge and Tovala smart oven. For budding chefs, there's the ChopBox smart cutting board with scales and SmartyPans, which tracks weight and temperature for optimized cooking with an app that makes suggestions.

INDUSTRIAL IOT (IIOT)

GE's Chief Economist Marco Annunziata said the Industrial

Internet brings together "intelligent machines, advanced analytics, and the creativity of people at work. It's the marriage of minds and machines." There is a growing Industrial IoT (*IIoT*) sector focused on improving diverse areas such as manufacturing and logistics, agriculture, and military. Following an estimated growth rate of approximately 30 percent, the IIoT market may be worth $14 trillion by the year 2030.

An easy place to envision the value of IIoT is in the manufacture of goods. Large machines used to create products at scale are expensive to buy and maintain, and unexpected breakdowns can cause a myriad of problems, from downtime to late orders to poor quality output. Machine health—sometimes measured as Overall Equipment Effectiveness (OEE)—is an easy domain for IIoT, where attached sensors provide various metrics on each machine. Providing measurements about, for example, the heat of a device falling outside of acceptable range can provide information for maintenance crews to be more reactive to breakdowns by tapping temperature data logs. Over time though, these devices can become smarter and begin alerting operations teams to more preventative measures, and eventually, they can proactively, predictively, and prescriptively solve breakdowns before they occur.

IIoT is also finding a perfect home in supply chains between the factory and retail floor. Historically starved of data and run by calendars and paper documents, global logistics are improving with real-time tracking of orders. Led by providers like Maersk, the ability to follow individual shipping containers with GPS in real time across the open ocean helps enterprises track orders with more fidelity. Taking this further, companies like CargoSense provide IIoT devices that anyone can add to their shipments. These devices can collect metrics like temperature and moisture levels, and report whether or not they stay within expected limits. They function as digital seals, validating that cargo was not tampered with. Such devices in logistics can help ensure quality, security, and compliance.

Due to their relatively low cost at scale, IIoT devices can be placed anywhere they're needed. They can act as auxiliary security by monitoring the grounds of a warehouse at night with hyperspectral cameras, provide early flood prediction by detecting ground moisture, or alert management to an empty break-room snack machine. The CalAmp and Pallet Alliance IoT wooden pallets can offer logistical transparency to ensure low costs without fundamental changes to existing processes.

Finally, in the retail space, POS systems are getting the gong. Consumers are able to make payments from mobile phones via NFC systems like Apple or Samsung Pay, and retail workers are replacing large POS systems with handheld scanners and mobile credit card Square devices. For the larger baskets of grocery stores, Caper's Smart Cart's scan items for checkout as you place them into the basket. But even these are stopgap measures as outfits transition to more flexible self-service shopping systems, like the entirely touchless camera-based system of Amazon Go, which we'll revisit soon. For now, let's step away from retail and venture into the greater world of smart cities.

SMART CITIES, SMART GRIDS, SMART WORLD

Unlike most social human endeavors, like nations or corporations, cities rarely die. Even Hiroshima, once decimated by a nuclear strike, is a thriving city again. Geoffrey West, researcher in complexity science and cofounder of the august Santa Fe Institute, succinctly explained why: "Adaptation, not equilibrium, is the rule." Cities are powerful because they are complex adaptive systems. Up until now, most of that adaptation has been driven by human intelligence and labor, but now we have a new kind of adaptive component. Projects that leverage IoT at scale to reduce costs and improve sustainability and other aspects of city life are called *smart cities* and account for nearly a quarter of all IoT projects. Smart cities aren't necessarily planned cities like Masdar in Abu Dhabi, but instead incorporate IoT devices to improve critical systems such as transportation, energy consumption, and safety. One of the first smart cities was the retrofitted ancient city of Seoul, South Korea.

Smart cities follow three phases of maturity. The first phase is *improving isolated operations* in the city, like traffic, environment, and culture. The second phase is *vertically integrating services* to improve processes via IoT, like how Palo Alto outfitted its traffic lights with an artificial intelligence mesh network, allowing them to coordinate with shifting traffic patterns. The third phase integrates the vertical systems, *providing holistic improvements across various service areas*. In Seoul's smart city experiment, this third phase meant all of the information worked together and was accessible through apps to make the people *smart users*. While cities will grow at different rates, it's difficult if not impossible to skip phases.

For cities to genuinely be smart, however, they must be sustainable—especially in relation to the electricity and power they consume. In the US and other countries, electrical transmission is generally designed through a brittle collection of localized power generation and usage. In 2021, the primary power generators in the world were still coal burning plants, which are also major contributors to global climate change. While only a few locations in the world have mineable coal, it can be transported just about anywhere to be burned and thus provide local power generation. On the other hand, low carbon footprint renewable energies like geothermal, solar, and wind can only be generated in certain places. The energy itself must be transported elsewhere, which is historically difficult to do.

In short, if we're ever to see majority usage of renewable energy, a smart grid that relies heavily on IoT will be required to track, manage, and transport electricity in an optimal and even predictive way. GE is a leader in this field through the Industrial Internet Consortium and has made massive investments in IIoT for generators like windmills and smart grid tech. Beyond the transportation of energy, cities can be outfitted to use less electricity. Like the Nest for smart homes, we can optimize energy use at the corporate, city, and national levels. This will be the key to creating a sustainable circular economy.

If we zoom out from smart areas (like cities) to smart networks (like grids), we eventually reach the point where we want to be smarter about the world in general. There's the obvious extension of grid and city tech across international borders, which is simply a matter of scale, but more than that, there's a need to collect and share knowledge of the global ecosystem as a whole. At the US Defense Advanced Research Projects Agency (DARPA), John Waterston is leading a project called the Ocean of Things. Over the next few years, the project aims to drop 50,000 IoT sensors into the world's oceans, with more planned in the future. Considering that more of the surface of the earth is covered in water than land, this will give humanity a huge boost in knowledge about our planet. As for the terrestrial, the company FGR has built a solar-powered IoT and data communications systems to monitor phosphate mines in dangerous and remote desert regions. These devices are outfitted with temperature, moisture, and pressure gauges. This data is funneled into a cloud system and paired with ambient sources like weather data, compliments of Cumulocity IoT. With more robust tracking, the system can increase the availability of fertilizer and help stabilize global food crops with the help of IoT devices deployed to track the state of industrial farms.

COMBINATORIAL CREATIVITY AND UBIQUITOUS COMPUTING

"We must always change, renew, rejuvenate ourselves; otherwise, we harden."

—J. W. VAN GOETHE

A shady-looking man slinks into a store, eyes darting around, slipping objects off the shelf into some dark recesses of his trench coat. He sidles to the exit. A security guard yells to the man, smiles, and informs him he's forgotten his receipt. This scene was from an IBM commercial in the nineties. Now it exists, as predicted, in the form of Amazon Go, a concept store in Seattle where users link their Amazon account upon entering. When a consumer grabs items off the shelf and exits with their bounty, their Amazon account is charged. No checkout lane, no human interaction. I tested the limits of this store by attempting to confuse it through a combination of moving items around, to slipping items into my jacket, to exiting and reentering multiple times. No dice. It's the holy grail of retail shopping: correct, frictionless, and instant.

Amazon didn't create any new class of technologies to implement Go. It's a handful of off-the-shelf components like cameras, 2D codes, other sensors, and lots of AI. The ability to create something new from a collection of barely accessible technologies is the essence of deep tech. The next decade will see a Cambrian explosion of new technological uses thanks to IoT. All that's required is a handful of components and the combinatorial creativity to put a subset of them to use in novel combinations.

With saving money being the father of all invention, some of the best IoT projects exist to clone expensive capabilities. Professor and inventor Joshua Siegel created an early IoT device called Carduino, leveraging car and smartphone sensors to track car health telematics. The device can be programmed to use an accelerometer to track tire pressure or roll up the car windows if it detects rain. Or it simply lets you remote start the car over the internet. Many of these features are built into modern cars, but this device—a clever hack—can retrofit old cars for just tens of dollars. And there are many other such examples. During the coronavirus pandemic, for instance, Rice University created an open source ventilator—dubbed the ApolloBVM—to combat the shortage of ventilators. It was buildable for around $200 US with off-the-shelf actuators, controlled by a couple of IoT platforms called *Arduino*, held together by some 3D printed parts.

Through this burst of combinatorial creativity, we can work toward a world long envisioned by science-fiction writers called *ubiquitous computing*. The concept is that once we live our lives surrounded by smart objects that are capable of measuring and acting upon their environment and sharing that information with each other, humans can effortlessly interact with their world without consciously or specifically executing commands to a particular device. Rather than picking up a "remote control" to "turn on" a "television" and changing to a "channel" to see a particular "show," in a world of voice activation, cheap smart walls, and AI assistants, we can walk into any room and ask, "What's going on in Bali today?" and get live streaming video and stats from the closest relevant and capable device, be it on the wall or on smart glasses. When computing devices fade into the background, no longer fighting for attention, we'll have reached a state of technology design called Calm tech.

But before we can hope to live in a world of ubiquitous computing, we first need to build the technologies. Let's cover the technical basics.

TECHNOLOGY

The Internet of Things is a loose collection of banal technologies that you'd be forgiven for neglecting. They lack the sex appeal of artificial intelligence or the promised social disruption of blockchain or autonomous vehicles. The most seductive thing that can be said about IoT parallels Willie Sutton's answer when asked why he robbed banks ("Because that's where the money is"): IoT is where the data is. More than any other deep technology available today, IoT is already innovating how we live and work, and quietly rewriting the rules of industrial operations as a major player in what's being called *Industry 4.0*, because it turns out the Information Age really is all about data.

The distinction between smart watches, smartphones, tablets, and laptops is a matter of degree, where we arbitrarily draw lines for marketing purposes. Is a smart watch a general purpose or specialty device? It probably depends on how smart it is, and even then, the lines are blurred. To channel Descartes, at what point does a laptop become a desktop and a desktop become a server? Industrial IoT devices are very much little computers with industrial-quality sensors attached. Cloud computers, too, are a valuable component of IoT infrastructure, as well as anything run on or accessible to them, from AI to blockchain to quantum computers. Wearables like smart watches fit the definition of IoT, but so do virtual and augmented reality devices. Autonomous vehicles are basically just multiple IoT devices on wheels. While many of the technologies named in this paragraph are not IoT per se, IoT is where they all come together, like at a cocktail party to mix and mingle.

To understand why IoT plays such a central role is to understand that they're general purpose in nature. If we attempted to list the myriad ways in which IoT could be put to use (an exhaustive list of potential use cases), we'd never finish. Instead, we'll keep this high level by dividing the Internet of Things into two major components: internet and things.

THE INTERNET

"The rich are different from you and me," claimed F. Scott Fitzgerald. "Yes," Ernest Hemingway replied, "they have more money." This amusing point belies the complexity of the difference. The same can be said for the Internet of Things: they're like normal things, they also just have the internet. But this distinction hardly suffices to explain the vast gulf between things that have the internet and things that don't. We'll cover how "things" are "interneted" in this section.

The idea of transmitting data between electronic devices is an old one, predated by nearly a century with Samuel Morse's telegraph and its alphanumeric encoding of *dits* and *dahs*. By modern standards, memorizing Morse code and tapping out long sentences feels tedious, but it beat the Pony Express in terms of speed. The Internet eventually allowed a network of research computers to transmit files with a series of ones and zeros, encoding signals much more powerfully than Morse. The first internet message was sent over half a century ago with the simple message "LO" (they attempted to type "LOGIN," but the system crashed part way through). Today, IoT devices leverage this original network protocol for uses as varied as streaming real-time data about foot pressure on a shoe to remotely reserving a taxi without speaking to a human. It turns out that the protocol was so general, it's still in use today and continues to be leveraged in unexpected ways.

It's worth looking into what makes the internet so powerful by diving into its constituent parts—a network of layers and topologies upon which the flexibility to power IoT is built. The following sections can be dense and technical, but understanding these components is integral to grasping the real power of IoT and what can be done with it.

ISO 7 Layers

The fundamental structure of the internet and much of its power lies in its protocols. The Open Systems Interconnection (OSI) model describes seven layers to a network: physical, data link, network, transport, session, presentation, and application. With these layers, you can design any kind of encoded transmission network, from transferring images between phones via Bluetooth to Voice Over IP (VOIP). The trick is getting others to agree to your proposals, but for our purposes, we'll concern ourselves with mainstream and emerging standards.

It all starts with physics in a first layer. Whether this first layer is wired or wireless, we need to physically transport some symbols from one place to another. A symbol, in this case, is a wave function of some physical process like radio waves or electrons. These waves pulse high and low, representing ones and zeros (aka binary digits, or bits) over the physical medium. This is leveraged by the second layer, called data link, which uses the physical transmission medium to reliably bundle up symbols into a data frame and moves them over the network between two devices.

We can support more than two devices in the network layer, or the third layer, where multiple devices (also known as nodes) and networks are connected to each other, and how packets of data route between them. This layer also determines our ability to find other computers on the network, like with Internet Protocol (IP) addresses. This layer, for example, is what lets your iPhone connect to a cell tower and eventually connect to Netflix. Now that we have a multinode network, we need to reliably send data across that network in a standard way.

The fourth layer is the transport layer, which is responsible for ensuring that segments of data are reliably transmitted to each other across the network. The internet that we generally think of, with web pages and streaming videos, largely uses the Transmission Control Protocol (TCP) for this layer. The internet is somewhat defined by layers three and four, and the protocol is often shortened to simply be TCP/IP (though other protocols certainly exist). This is a good example of how several OSI layers can be so tightly linked that they blur together.

Next, networks start feeling more concrete. With the physical ability to transmit bits and the ability to route those bits in a meaningful and reliable way to another device found on the other side of the globe, the top layers are what give meaning to these ones and zeros zipping around. Layer five is the session layer, which is chock-full of technical geekery, but in essence, it allows you to have long-running, continuous transmissions between nodes, like a phone call.

Layer six is the presentation or syntax layer, which is what gives meaning to all of the blocks of data we've been keeping as binary digits. The presentation layer is what defines a string of bits to be an image, a video, or an emoji. The final top layer, the seventh application layer, is where the user interacts with the network. The application layer is where we might view a web page using the hypertext transport protocol (more commonly known as HTTP) or log into a remote command line via secure shell (SSH). Layers one through six are made for computers, and layer seven is made for us and computers.

We must also consider more than just the layers of the network. We have to think about how nodes are laid out as network topologies (peer to peer, star, mesh), consistency versus availability versus latency tradeoffs, risk concerns like protocol popularity, chattiness versus battery drain, and conceptual concerns like openness. The answers to these considerations are represented by a veritable zoo of protocols.

NETWORK HIERARCHIES AND TOPOLOGIES

While the OSI layers are a generalization of the components involved in building a robust data network, we need to know a bit more before we dive into exactly which IoT network protocols are useful for our needs. One is an understanding of how our devices are configured to communicate with one another, also known as the *network topology*. Does our smart thermostat connect to a smart home hub alongside our other home devices, or do they connect with each other in a mesh? Does my smart watch have an LTE chip allowing it to connect directly to the nearest cell tower, or does it use Bluetooth to connect to my smartphone, which in turn provides internet services via the tower? Answers to these kinds of questions are driven by the type of user experience one wants to provide and will drive the sort of networking protocols they need to support.

Another piece of the puzzle is understanding the range the devices need to operate in. An industrial data collection device on a mill machine need only communicate within the four walls of the factory that houses it. Meanwhile, an Arctic barometric device may need to send readings to a base station several miles away, or perhaps even to an Earth-orbiting satellite. Luckily for us, there's a well-understood hierarchy of IoT network types based on the range or wireless communication.

The smallest range is the nanonetwork, which is the communication of microscopic electronic elements over millimeter ranges. The next level is the kind of Near-Field Communication that allows touchless cards to unlock hotel rooms, or how your phone communicates via Samsung or Apple Pay. Then there are Body Area Networks (BAN) for wearables and Personal Area Networks (PAN) for your immediate workspace, such as your Bluetooth mouse at work. Next are the more common Local Area Network (LAN), often Wi-Fi-connected devices like smart TVs in your home or laptops at work. Broader still are Campus Area Networks (CAN), Municipal Area Networks (MAN), and finally, Wide Area Networks (WAN) such as the globally connected internet.

A question every IoT device maker needs to ask is: "How broadly does my device need to communicate?" An Apple Watch may only need Bluetooth to connect to a smartphone as a PAN, while smart parking garage tracking devices could be connected via LAN or CAN.

Now that we've covered ISO and hierarchies, let's take a break from dense concepts and technical jargon and look at a few examples of IoT-focused network protocols in use today, along with their strengths and weaknesses.

Wireless Protocol Suites

"When wireless is perfectly applied the whole earth will be converted into a huge brain, which in fact it is, all things being particles of a real and rhythmic whole. We shall be able to communicate with one another instantly, irrespective of distance. Not only this, but through television and telephony we shall see and hear one another as perfectly as though we were face to face, despite intervening distances of thousands of miles; and the instruments through which we shall be able to do this will be amazingly simple compared with our present telephone. A man will be able to carry one in his vest pocket."

—NIKOLA TESLA, 1926

There's a relationship between power consumption and network traffic. The general rule is, the larger the volume of data transmitted, the more power consumed. Think of a shovel for dirt. A big shovel holds more dirt per motion but also requires more strength to lift it. But a small shovel used rapidly is also exhausting. You can't circumvent the fact that the more dirt you move, the more energy is required, but you *can* design shovels that are optimized for certain jobs. In wireless electronics, balancing energy consumption and data transmission is an endless struggle. There are several competing low-power wireless transmission standards vying for supremacy in the IoT space, each corresponding to different layers of the OSI model.

When trying to decide upon the correct network protocols, a

good rule of thumb is to start with your physical range requirements. If this is a wearable, you need near field; if it's a parking garage IoT, you may need wide area. From there, aim for the most mainstream protocols (NFC, Wi-Fi) unless there's a technical reason (power usage, data rate, range, etc.) to land on a less common network (Neul or LoRaWAN). Next, bias toward open protocols, unless there's a business or technical reason why a proprietary protocol better fits your needs (Thread or ZigBee).

While IoT is often described in terms of the physical devices themselves, it's important to note that the complexity of connecting those devices to the internet is a substantial portion of IoT considerations. As a friend in the IoT space is fond of saying: "There are no cell towers in the ocean." Choosing how to connect IoT to the internet is often a decision made based on circumstances. The complexity is in understanding the options and decision criteria. Many breakdowns can be found online and in books, and some are referenced at the end of this chapter. Now, let's move on to the other half of IoT: the "things."

THE THINGS

"All parts should go together without forcing. You must remember that the parts you are reassembling were disassembled by you. Therefore, if you can't get them together again, there must be a reason. By all means, do not use a hammer."

—IBM MANUAL, 1925

There are many ways to break apart the physical components that make up "things" in the IoT world, but a focus on the following three can be a useful division: transducers, computers, and power supplies. Each of the three contain deeper constituent parts, but in practice, these are good bailiwicks of concern.

TRANSDUCERS: SENSORS AND ACTUATORS

Transducers come in two flavors: sensors and actuators. Sensors, as their namesake implies, sense the world around them. Your nerves are sensors that convert various signals into impulses your brain can understand, such as light into color or heat into feeling. Digital sensors act in much the same way but rather convert analog signals to digital codes that a computer can understand.

The popular Apple Watch packs in a handful of analog and digital sensors: compass, global navigation, altimeter, electrical and optical heart sensors, accelerometer, gyroscope, ambient light sensor, microphone, and pressure sensor. Most of these sensors are small combination chips, like the compass, accelerometer, and gyroscope. Or a single chip supports multiple standards, like global navigation that works with GPS, GLONASS, Galileo, and QZSS. A common use of IoT devices are simply to collect data, from weather sensors to smart home scales. The active ingredients of these devices are mainly sensors connected to the internet.

Actuators are, in many ways, the opposite of sensors. While sensors are responsible for inputs into the computer, actuators are how a computing device interacts with the outside world. Or in other terms, sensors are for inputs, and actuators are for outputs. In our Apple Watch example, actuators would be the watch's speaker, OLED display, and haptic feedback system. Even the simplest IoT device, like a smart TV, is more than a screen. It contains IR/RF readers for remote controls, as well as commonly ambient light sensors to dim in the dark. Few IoT devices exist as actuators alone.

COMPUTERS, STORAGE, AND COMMUNICATION

The guts of IoT devices are just tiny computers. Boring old little computers made up of microprocessors for computation, memory like RAM, longer-term storage like NAND flash, and data transmission like a Wi-Fi shield. Not all of these parts are necessary, but they're common for an IoT computer. What makes most IoT computer hardware interesting is that designers and manufacturers have shrunk them down to unfathomable sizes, reducing power consumption dramatically, and selling devices at scale for stupefyingly low prices.

The need for IoT computers is so common, they're sold as platforms for targeted use cases. For example, Arduino is the granddaddy of open microcontrollers, while Raspberry PI is a more full-fledged computer running an operating system and various onboard components. These are also known as a system on a chip (SoC). While you'd use an Arduino for managing a collection of transducers for simple inputs and outputs at low power, a Raspberry Pi is more suitable for complex processing, such as running an AI for processing images collected by a connected camera.

In 2019, the Apple Watch contained an S5 chip with a 64-bit dual-core processor and W3 chips. Between the two, in addition to computation, they were also responsible for wireless data transmission, supporting cellular phone calls (LTE and UMTS3), Wi-Fi (802.11b/g/n 2.4GHz), BLE (Bluetooth 5.0), and NFC for Apple Pay. Whether an IoT device heavily leverages sensors to read the world or actuators to interact with it, one component all IoT devices have in common is a microcontroller. Even passive RFID tags contain a chip for the simplest processing.

POWER SUPPLY: WE MEAN BATTERIES

Relative to other computing components like CPUs, batteries are huge, heavy, and slow to improve. Most of the tradeoffs you'll make when building IoT devices are related to shrinking power consumption, therefore shrinking battery size and charging requirements. The tradeoffs are endless and frustrating, from slower onboard computation to shorter wireless data transfer range, from less frequent sensor reading to simply removing components.

Unlike Moore's Law, which says microchip transistors double every two years or so, battery technology capacity tends to double only every twenty years. Nowhere is this more apparent than when you tear apart a smartphone. Once you remove the casing and display, you'll find that most of the innards are comprised of a battery with tiny adjacent computers and sensors, which accounts for their increased battery life over the years.

Batteries are the bulk of modern IoT devices.

We don't need to be experts in Kirchoff's circuit laws to consider the high-level business tradeoffs between the data we need to transfer at a minimum, how small we need the devices to be, and how long we want the device to last between charges. Some companies are getting creative. Wiliot is a company building Bluetooth chips powered by batteries, but these batteries are charged by harvesting ambient radiation like Wi-Fi or FM radio. In theory, they can run endlessly as long as there's some radiation nearby. The concept is not dissimilar from solar panels; these just operate without sunlight.

Cleverness and creativity have allowed the industry to make power improvements despite the shortcomings of batteries, but these aren't the only problems that impact IoT.

LINGERING PROBLEMS

Compared to many other emerging technologies like blockchain or quantum computing, IoT has few downsides or existential threats in its future. But nothing is perfect, and there are still structural concerns to address as we roll out this new tech.

CONFIDENCE

Do you ever forget to drink water? Consider HidrateSpark, a smart water bottle that "glows to make sure that you never forget to drink your water again," apparently preparing for a world where humans have lost their sense of thirst. What about Smalt, the internet-connected salt shaker, or Toasteroid to burn an image into your toast via a connected mobile app? Or how about Flatev, the Keurig-like machine that converts pods of dough into instant tortillas? We could go on. A unique story of trolling from the Consumer Electronics Symposium (CES) in 2020 was that of an IoT potato. It didn't do anything at all, it was merely a potato with an antenna sticking out of it, but it still had a line of gawkers curious about what it could do. This acted as exhibit A, showing that IoT can easily jump the shark.

Emerging technologies always have a precarious path to mass adoption, and a deluge of bad ideas can stall public confidence away from genuinely useful devices. And while it seems doubtful that any of these are a threat to eventual IoT adoption, a lack of creativity can sour customers and corporations away from valuable ideas and delay real benefit for years.

INTERNET AVAILABILITY

Without the internet, computerized things have only a fraction of their use. Although improving by the year, ubiquitous network access is far from being a solved problem. We need a collection of network options to support wider areas, lower power, higher availability, lower cost, and all of the other incremental improvements we expect from technology. Happily, a few projects are working to build the networks of the future.

Project Loon, from Alphabet's Google X division, is focused on releasing suborbital balloons high into Earth's atmosphere on the border of space to provide internet access to underserved communities in remote regions around the globe. Not to be outdone, Facebook *Aquila* has similar aspirations with solar-powered drones in continuous flight. For a more open network, Project OWL (Organization, Whereabouts, and Logistics), sponsored by the open source Linux Foundation, creates a mesh network of IoT devices called DuckLinks. OWL is designed to deploy networks quickly in disaster scenarios, but it can potentially play a more permanent role. Through these projects, along with a handful of satellite internet initiatives, every corner of the earth may have some form of network access within a generation. Pair these efforts with the relentless push by cell providers into higher bandwidth 5G, and it's obvious the future is trending toward higher speed and consistently available internet. This only increases the mobility of IoT devices, untethering them from homes and smartphones.

ENERGY

Energy is the voracious vampire sapping vigor from many promising projects. It's why Google Glass only survived about an hour of moderate continuous use. Tile, while great, still requires me to change the battery to find my luggage. Many in the industry rank power supply as a top concern in IoT product design.

Wireless power transfer has been attempted for over one hundred years and was a particular obsession of the great electricity pioneer Nikola Tesla. While there have been some improvements in wireless transmission of power within close contact, such as the Qi standard supported by most modern smartphones, longer range wireless power transfer still eludes us. uBeam and Energous are two famous attempts, and we shouldn't be hard on the myriad of imperfections in their approaches. MIT has made some recent progress with power over Wi-Fi, but motion energy transfer is currently a more practical option.

For now, the best we can do is find ways to require less power or to harvest power from the ambient environment. For example, automatic watches have been self-winding for centuries, converting movement into potential energy through the winding of a spring. JBL recently released low-power headphones—so low in fact, they can be charged with minimal solar. Taking this one step further, researchers at the University of Waterloo created a battery-free remote input device called Tip-Tap, which you wear on your fingers like a glove. Smart shoes from companies like SolePower collect electricity from mere walking. Pair these technologies with improvements like IBM Research Battery Lab's new experiments in battery elements extracted from seawater, and in a few decades, we may have this battery problem licked.

E-WASTE

Famed Russian writer Anton Chekhov once observed, "Only entropy comes easy." The aging of IoT devices will not resemble the slick veneer of fifties futurism. The ships of time will discard an endless jetsam of corrosion—part *Star Wars* Tatooine, part *Jetsons*. We call this refuse e-waste. Beyond the visual blight, many electronics and batteries contain rare earth metals, as well as volatile and carcinogenic chemicals that can poison animals (including humans), soil, water, and eventually, Earth's food supply. To avoid a future where we're all wading through electronic trash heaps, we have a few issues to solve. Luckily, any kid growing up in the middle of the twentieth century knows the solution: reduce, reuse, recycle.

The first step is reducing how many IoT devices we use. This means balancing our physical needs against how many things we want. Rather than buying a handful of devices to detect moisture levels in your basement, another to detect temperature, and another to detect movement, you're better off with a single device that can do all three. The benefit to the consumer, beyond less waste, is that multiuse devices cost less overall and require less frequent battery changes. The easiest example of a single popular multiuse device is a smartphone (which explains why so many other single-use personal devices have had a difficult time taking off).

The next focus is to reuse the IoT devices we have. There are many programs that send gently used electronics to others, such as Portland's Free Geek program. Beyond giving electronics a second life and reducing waste, these programs provide electronics to underserved communities who may have a harder time taking part in the IoT revolution.

The last step in dealing with e-waste is to recycle or reclaim IoT materials. Many rare earth metals are of limited supply—they can be extracted from devices and reused in new component manufacturing. Some cannot but should be reclaimed for the health of our planet. As IoT manufacturing scales up, we must match that growth by improving and scaling up recycling methods. Batteries alone contain important metals like lithium, manganese, copper, and cobalt, all of which must be repurposed at scale, lest we run out (at best) or they find their way into our oceans (at worst).

PRIVACY AND SECURITY

In 2015, the Chinese Ministry of Public Security announced a goal that shocked the world. They wanted to build a database of China's 1.3 billion citizens with the intent to track their movement and actions anywhere in the country within three seconds. They'd leverage AI, powered by a network of nationwide IoT cameras. By 2019, the system became fully functional at scale, while growing in speed and sophistication. Before other countries point fingers, the USA's PRISM program and UK's Tempora system are clear examples of Western snooping. But while technologies like big data and AI open the door to global digital surveillance, the last mile of data collection lies in IoT. Yes, the NSA can read your emails, but China wanted IoT to take mass surveillance into the real world.

Nation-state actors are not our only source of privacy and security concerns. Target's database of forty million credit and debit card accounts was famously stolen in late 2013 when their HVAC's IoT system was hacked. In 2017, an internet-connected fish tank was used to launch an attack on a nearby casino, stealing ten GB of data. A year later, parents everywhere were horrified to learn their Wi-Fi-enabled baby monitors were vulnerable to hacks, acting as remote spy cams for would-be thieves.

Security and privacy attacks are varied, weird, and increase every year. The IoT community is finally starting to take security seriously. The Internet of Things Security Foundation (IoTSF) and the Trusted IoT Alliance are some of the emerging groups dedicated to promoting tighter collaboration between device makers, users, and security experts. There is also a burgeoning field of IoT network security services, many leveraging AI to detect and track suspicious IoT network traffic, like Senr.io. Expertise in IoT security will be a busy field in this decade.

Beyond being open to vulnerabilities, IoT networks can also be weaponized. Political scientists call this phenomenon "dual-use technology," similar to how nuclear reactions can be used for both power generation and the building of bombs. Vast collections of IoT devices can be compromised to take down systems by something called a distributed denial of service (DDoS) attack. This is similar to how a single bee sting might hurt, but a swarm can be deadly. IoT devices can be compromised over several years, building a botnet, and in an instant, a state actor or even a single person can launch a coordinated attack at scale. In mid-2019, hackers released Russian Intelligence Service FSB documents describing IoT botnets that were under the country's control. IoT is yet another front in the endless fight for security.

COMPETING STANDARDS AND PROTOCOLS

Google Nest, Apple HomeKit, Amazon Alexa. ZigBee, Z-Wave, Thread. Industrial Internet Consortium (IIC), OneM2M, Object Management Group. These are a small taste of the myriad competing standards in the IoT space. Too many standards? There's also a standard for that. Several, in fact, be it Open Interconnect Consortium (OIC) backed by Intel, or AllSeen Alliance born from Qualcomm.

What's interesting in the IoT space is not a lack of strong partnerships or consortiums but an overabundance of them. This balkanization of IoT platform standards should not be cause for concern, however. Since most consumer home devices support multiple standards, this pattern is increasingly followed by industrial IoT. For example, my August Smart Lock supports Apple HomeKit, Amazon Alexa, Google Assistant, and more. There's also good precedent for the market to demand a common interoperable standard that allows consumers to use competitors on the same network. Movies Anywhere is a great example of this, where Amazon, Apple, Google, and Vudu have all agreed that movie purchases on one network could be streamed through any of the others. This consortium cements consumer belief that they actually own the purchased movie. The 2019 announcement by Apple, Google, and Amazon to align their platforms into a unified standard is good news for users of those services but bad news for anyone wishing to break into the oligopoly.

All of these problems aside, the good news is that none of them are inherently unsolvable. Each of these problems presents an opportunity to make things better, not only for the industries already engaged in IoT, but also for anyone who wants to solve problems in the burgeoning IoT industry. With seventy-five devices per person by 2030, there are plenty of users to serve.

IOT PEAK END

"I for one welcome our new robot overlords."

—KEN JENNINGS, *JEOPARDY* CHAMPION

The Internet of Things will be a complete and forever shift in how humans interact with the world. Once we all get used to smart things, we'll be loath to accept dumb ones, once price ceases to be an issue. Smart will become the new normal, and basic refrigerators will go the way of hand-crank car windows and TV knobs. Through wearables, smart homes, and offices, much of our world will be understandable, controllable, and automatable at ever deeper levels.

Anything prefixed with "smart" is likely IoT, but things can get even smarter. IoT will, in many ways, also be many people's first direct experience with artificial intelligence. While AI algorithms have been predicting weather and curating personalized media for years, it's a vague thing in the background and is considerably less direct than chatting with Alexa on an Amazon Echo IoT speaker. And with a wide collection of IoT devices interacting with aspects of an environment, it will become the frontline technology of what Microsoft's Satya Nadella calls the "intelligent edge." It's one thing to have IoT smart fabrics; it's another thing to have it intelligently reconfigure a garment based on expected rain.

As Google chairman Eric Schmidt said on a panel at the World Economic Forum:

> The Internet will disappear. There will be so many IP addresses, so many devices, sensors, things that you are wearing, things that you are interacting with, that you won't even sense it. It will be part of your presence all the time. Imagine you walk into a room, and the room is dynamic...you are interacting with the things going on in the room.

Mass adoption of IoT is one big step toward the AI of things, and eventually, the full ubiquitous computing described by Schmidt.

We're growing used to a world where weather predictions are always available at our fingertips through smartphones. A global net of worldwide IoT sensors is a necessary step in such desires and expectations. With grander visions still, we want to save endangered species, reduce rain forest destruction, and keep pollution from our water supplies. Pure donut economics. Leveraging the power of IoT, anyone can play a major part in helping humanity live harmoniously with our world, ensuring we thrive within acceptable boundaries.

The realization of the many IoT visions are building an unprecedented opportunity for organizations and ambitious individuals. Of all the deep tech in this book, IoT alone is expected to add $15 trillion to the annual gross world product in the year 2030, give or take a couple trillion. This global network will only drive the need for more. More accuracy, more granularity, more precision. More opportunity. More IoT.

FURTHER READING

- *IoT Inc: How Your Company Can Use the Internet of Things to Win in the Outcome Economy.* Bruce Sinclair
- *The Internet of Things.* Samuel Greengard
- *Calm Technology: Principles and Patterns for Non-Intrusive Design.* Amber Case
- *Building Wireless Sensor Networks: With ZigBee, XBee, Arduino, and Processing.* Robert Faludi

6

THE ONCE AND FUTURE AV

"No matter where you go, there you are."

—BUCKAROO BANZAI

Attendees at the 1939 World's Fair Futurama Exhibit.

In the beginning, humans hunted and gathered. We lived in small, roving tribes, fighting the elements and each other, all the while discovering fire, cooking, clothing, and the wheel. Eventually, we settled into agrarian societies and made one of the greatest discoveries in history: semiautonomous transportation. These vehicles helped us farm more effectively, maneuver quickly, and better battle each other. They required only minor supervision once properly trained and were hesitant to put themselves in danger, which meant they looked out for the security of their human passengers. As a whole, horses, oxen, and other beasts of burden were pretty good. Riding on the backs of this animal technology, humankind rose to master the earth.

Then, in 1885, we ruined everything by inventing the internal-combustion powered automobile. Suddenly, transportation required constant human engagement, lest the equipment run into obvious objects like trees or pedestrians. Furthermore, these new vehicles consumed processed fossilized hydrocarbons and emitted global death gas, but they moved faster than camels. One hundred and thirty years later, we're on the brink of converging the speed and power of mechanical vehicles with the self-preservation and learning capability of beasts. We're close to mastering autonomous vehicles.

The invention of autonomous vehicles (AV) capable of driving themselves from point A to point B without human intervention is one of the most disruptive technologies of the twenty-first century, but it's hardly a new idea. As automobiles proliferated and replaced horses in city traffic, people realized that driving is a tedious and dangerous activity. In 1939, General Motors sponsored a World's Fair exhibit, called Futurama, purported to peek into the year 1959. Although in modern times, we'd find the show mundane—it focused on the value of an interstate highway system and single-direction roads for high-speed travel—it did demonstrate a vision of a future with an automated highway system, purported to fix issues like city pollution and the growing death toll from automobile accidents. Today, we speak of autonomous vehicles as a future technology, but long ago it was thought to be a mere twenty years away. We're sixty years late.

While generations dreamed of cars that drove themselves, DARPA was thinking about warfare. A country's ability to win in war has always been related to its ability to transport troops and equipment over vast distances. Rome secured much of the world with the most complex network of roads in the ancient world, the British Empire kept control over waterways through its sophisticated Navy, and the United States built a domestic interstate highway system as a defensive measure for the Cold War. In the early 2000s, an emerging problem of improvised explosive devices plagued coalition forces. These cheap and easy explosives disrupted transport, killing and wounding many soldiers. DARPA, in search of a solution, decided to employ the best roboticists in America to help invent automated long-distance transport. They did this by staging a competition called the DARPA Grand Challenge—a vehicular race through the desert with the goal of completing the course without human intervention. In order to win the $1 million purse, the vehicles had to be autonomous.

The first challenge, held in the Mojave Desert in March of 2004, yielded no winners. After two years of development, the top team only completed seven miles of the 150-mile course. Undeterred, DARPA held a second race in October of 2005. This time, five teams completed a 132-mile course. A team from Stanford University won, whose AV completed the course in just under seven hours, with an average speed of about nineteen miles per hour. DARPA went on to host more autonomous challenges. They achieved their goal, made a spark, and the world realized that AVs were possible.

Although there was a long way before the commercial viability of AVs, many who took part in the DARPA challenges found ways to apply the skills they learned. One such person was engineer Anthony Levandowski, who, after building an autonomous motorcycle for the first Grand Challenge, later captured the public's attention by building a self-driving Prius that delivered a pizza across the Bay Bridge in San Francisco. Levandowski partnered with the second Grand Challenge winner, Stanford Professor Sebastian Thrun. Together, they founded Google's secretive AV Project Chauffeur, which, a few years later, became a multi-billion-dollar commercial venture called Waymo. Over the past decade, other companies have joined the fray—Cruise backed by GM, Zoox backed by Amazon, and even Apple started work on AVs. For the first time in history, we're at the cusp of reliable, large-scale, small-batch transportation without human intervention, unbound by the constraints of dedicated infrastructure like train tracks or commercial constraints like large populations of people.

Let's look at the benefits AVs provide (such as reducing casualties caused by imperfect human drivers), some of the technology behind how they work, why the time is now, and some of the problems and opportunities left to solve.

LIVES AND OTHER BENEFITS

To focus exclusively on autonomous human transportation vehicles misses much of the revolution that's happening right now. Pit mining equipment is moving toward full automation. Electric vehicles are disrupting the automotive and fuel industry and related supply chains. Ride-sharing is changing the culture of car ownership and, by extension, laws and city planning. AVs will only accelerate this change. Autonomous ground and aerial package transport drones are improving last-mile package delivery (i.e. FedEx and Amazon). The physically impaired will gain the freedom of independent mobility. Autonomous vehicles will save lives. AV makes way for more possibilities than what may appear at first blush.

SAVING LIVES

Automobiles are involved in the deaths of an estimated 1.3 million people per year, and twenty times more than that are injured. In the modern age, this figure surpasses deaths from wars, drugs, and violent crime combined. Automobile accidents are the number one killer of teenagers and young adults. In fact, five people were injured or killed in an automobile accident in the time it took for you to read this single sentence. If any disease or military attacks facilitated such high death tolls, we'd hold endless vigils, and politicians would run on platforms calling for their eradication.

But there may be a cure. The problem of allowing distractible, simian-brained humans to operate vehicles is a vincible one. Humankind has a long history of removing people from dangerous or hazardous jobs and replacing them with machines. The advent of autonomous vehicles is just another case study in our ingenuity.

The average driver might clear over 800,000 miles in a lifetime and be involved in an accident once every 200,000 miles. An intelligent digital driver, trained with the combined experience of an entire fleet of cars could rack up billions of miles in the real world, trillions in simulations, and participate in more rare situations than any individual in history. Every year, the brains behind autonomous vehicles get smarter, and unlike every fifteen- or sixteen-year-old with a learner's permit, they don't need to start from scratch with each new driver. Every new accident or novel situation makes the whole system smarter.

The system continuously trains an AI model and periodically deploys updates to the entire fleet. Self-driving cars commoditize transportation.

Without any other benefits, AVs should exist for this reason alone. And like any general purpose technology, autonomous vehicles can add more.

SELF-DRIVING CARS

Of all the use cases for autonomous vehicles, personal transportation is the most straightforward. It's easy to imagine a robot chauffeur. I rode with a model driver and found the experience to be like tooling around as the passenger of a cautious nanny. It cruised slightly below the speed limit, kept plenty of distance between cars, avoided lane changes, and signaled every turn. When we first hear of AVs that move, steer, and navigate automatically, most of our minds drift to self-driving cars (SDC). Those of us who commute to work can't help but daydream of being in an SDC while stuck in rush-hour traffic, but there's not much to say about the experience. If you've ever ridden in public transport, you can do the same tasks in an SDC but in private and from door to door. You can check email, watch videos, and chitchat. The more adventurous may sleep or even drink.

Many of the benefits of AV technology as applied to create SDCs is already being realized today. What SDCs will do is make these services profitable or cheaper. For this reason, ride-sharing companies like Uber and Lyft are having a hard time turning a profit in the early 2020s and are still burning through investment dollars. These business models will increasingly rely on SDCs to compete with the low cost of self-car ownership or existing alternatives like public transportation or taxis long term. AVs and ride-sharing combined can shrink the cost of personal transport from an estimated $1.50 to 25¢ per mile. Such savings will drastically reduce the incentives of car ownership in favor of fractional payments based on per-use criteria, such as number of trips, distance, or timelines. Picture a few local community AVs maintained by a fleet company, rented out to dozens of neighbors at a low cost. The neighborhood conflict of the future may not be jealously of the one who owns a luxury car but of the jerk who overuses the shared convertible.

Another emerging value of AVs centers around the concept of personal logistics. Out of necessity, the 2020 pandemic forced companies to speed up adoption of what's known as BOPIS (buy online, pick-up in store). Uber CEO Dara Khosrowshahi has said, "The COVID crisis has moved delivery from a luxury to a utility." Major retailers like Target and Home Depot created separate parking spaces for humans who picked up online orders. It's a short leap to optimize this infrastructure a bit more: simply remove the human driver. Suddenly, last-mile delivery of groceries and sundries can be inverted. Everyone can just send their car to pick up on demand, rather than waiting for the Amazon driver to make her rounds for the day.

The ability to automatically commute from anywhere to anywhere will change where people choose to live, potentially in a renewed suburban flight. The skyline may change a bit as well. With fewer eyes on the road, landscape advertising, like billboards, start to lose their value. These are minor examples. What other infrastructure changes may we expect?

City Planning and Real Estate

What makes the promise of SDCs so interesting is how society and the landscape will change when adoption happens en masse. The most direct changes will happen to the roads themselves. Roads are measured in area. The area of a good road should support the peak number of vehicles times the size of vehicles, plus human reaction time based on speed. If it takes two car lengths for a driver to react and stop, then that driver needs to take up three car lengths worth of pavement on the road. According to simulations run by Budapest University of Technology and Economics, full adoption of AVs will mark a 20 percent improvement in traffic flow and an even higher increase in traffic density. SDCs can react more quickly than humans, thus reducing reaction times. They can also map and change course in real time. Goodbye new ten-lane highways, hello high-density, single-file car lines—a high speed ballet of precision. Imagine your city or town with 20 percent less road, with no disruption in your ability to get where you want to go.

When roads are AV-only, the width of roads can change as well. Skinny lanes that cars navigate effortlessly can free up room for an additional lane, perhaps a high-occupancy vehicle (HOV) lane to support carpooling. Or better yet, create room for more bicycles, pedestrians, retail spaces, or outdoor restaurant seating.

In many cities, around one-third of all usable space is dedicated to parking. In a world where shared AVs are generally on the move—dropping people off and immediately picking up a package delivery—much of that space can be reclaimed. This new availability of valuable space could shrink the cost of real estate in city cores or grow the amount of green space. Consider the High Line in New York City, a former elevated train track that was converted to a 1.5 mile park. Much of this new real estate can be replaced, reclaimed, or sold. Useable commutes—commutes you can devote to things other than driving—will alter how far people are willing to travel to work. People may decide to live farther away, further disrupting city real estate.

As SDCs become the dominant form of personal transport, much of the infrastructure of traffic management can disappear as well. AVs are efficient drivers. Pair that efficiency with IoT transmitters that allow vehicles to communicate with other AVs at the speed of light, and we'll find that AVs will be able to negotiate their intention to turn or drive through an intersection. When the speed limit is codified in a shared digital map that all AVs can access, traffic signs and lights will become unnecessary. Why bother with speed cameras or traffic cops when cars are programmed to never go above a legal limit? Removing the need to maintain all of this infrastructure will save cities millions of dollars each year. These savings could balance out the massive loss in municipal fees from parking and speeding tickets.

After over a century of trying, we may finally be on the cusp of cities and towns realizing many of the benefits of mass transit technology: low cost, low footprint, and high speed, but fundamentally designed to go door to door.

INDUSTRY 4.0

In the same way our ancestors called automobiles "horseless carriages," limiting AVs simply as self-driving cars narrows our vision around their potential impact. AVs are more than Uber with a robot chauffeur. The ability to move objects without human intervention will change all manner of transportation tasks, most notably in industry and logistics.

The first large-scale use of AVs was not for human transportation but heavy mining equipment. The Rio Tinto Group built an autonomous mine in 2008 in the Pilbara region in Australia, which they monitored remotely from Perth, 1,000 kilometers away. Caterpillar's *MineStar* AV is generally available and operating globally. Another early adoption was autonomous tractors for harvesting large-scale farms. Like mining, this was an easier use case since the range is bounded. In 2015, a convergence of AI, robotics, and AV birthed *Greenbot*. This helpful little fellow is capable of functioning in smaller outdoor areas for a variety of general uses, from mowing lawns to picking fruit.

Like many other technologies that require the budget and support of government, unmanned ground vehicles (UGVs) have existed in militaries for eighty years, since the first Russian radio controlled teletanks. As we know, much of the modern commercial work in AVs was spurred by the three-million-dollar purse of DARPA's self-driving Grand Challenges. While the purpose of the prize was to promote research for military improvements to UGVs, AVs have grown into the burgeoning industry we see today. AVs are one of the more successful innovations to come from government R&D investment, alongside the internet and Tang.

AVs also have the potential for massive disruption in Industry 4.0 supply chain operations. Think of Amazon-style automated warehouses powered by thousands of AV picker robots, or Honeywell's robots loading Embark's autonomous long-haul trucks. Couple this with the strides made in last-mile delivery to leverage smaller personal AV delivery systems, such as Uber and drones, and it's evident we're entering an age of fully automated logistics.

ENVIRONMENTAL IMPACTS

We are now firmly in the Anthropocene Era, where humans are the dominant factor in affecting the global climate. We're experiencing record carbon dioxide parts per million (CO_2 PPM) and routinely breaking the hottest temperatures on record. This increased instability in the atmosphere results in stronger storms, bigger floods, longer droughts, and rising extinctions. All signs point to rising ocean levels, which will displace humans living in coastal regions and islands.

Today, the average car pumps six tons of carbon dioxide into the atmosphere every year, a major contributor to global climate change. An equivalent electric car, charged by connecting to a power grid, creates between two and four tons of CO_2. The generation of electricity still releases CO_2 into the atmosphere, and so has been dubbed the "long tailpipe." Moreover, the act of manufacturing an electric vehicle requires more carbon and rare earth metals than the manufacture of an internal combustion engine car. On balance, electric cars still come out on top by reducing emissions by up to 50 percent, depending on the lifetime of the vehicle. A third technology, hydrogen fuel cell vehicles, sit somewhere between the two. The good news is that the transportation industry is slowly tackling climate change, but we can do better.

New fuels are only part of the equation for reducing CO_2 emissions.

Brian Johnson, an analyst from Barclays, estimates that car sales will plummet 40 percent over the next twenty-five years as the use of rent-per-use fleet vehicles powered by AVs increases. Reducing the manufacture of excess vehicles will further reduce the carbon footprint of cars that normally spend 95 percent of their lifetime sitting idle.

Moreover, autonomous vehicles are simply better, more fuel-efficient drivers. There is a subculture of people called *hypermilers*, who optimize their driving through a series of techniques like scientific route planning, slower driving, or drafting from other vehicles to minimize fuel consumption. Cars driven by AI can continually improve these positive driving habits at scale. Think of it this way: your daily commute may require a gallon of gas, but an AV might optimize the same trip with half a tank, saving you in fuel costs and reducing carbon emissions.

Taken all together, new technologies and AVs could reduce annual CO_2 emissions from transportation by about 6.5 billion tons, a 17 percent reduction of global human carbon output. Only about one-fifth of those savings come from alternative fuels; the rest are from the proliferation of AV fleets. The effectiveness of automation to fight climate change in the transportation space will dwarf any particular switch from gas to electric vehicles (EV) or hydrogen vehicles (HV). While alternative fuels are worth pursuing for a number of reasons, if governments are truly concerned about the volume of atmospheric carbon, the best transportation investment they can make is funding the adoption of AV fleets.

TECHNOLOGY

In 2020, Waymo is the quintessence of AV companies, with a car that looks like a robotic marshmallow topped by an obsidian gumdrop. It's polished and clean with utopian futurism, but the idea that Waymo and other AVs represent is quite old. As mentioned, the World's Fair Futurama exhibit believed AVs would exist by 1960. Yet, despite the obvious benefits, AVs never took hold.

THE SECRET SAUCE IS AI

The missing component to making AVs operational was artificial intelligence. While many technologies have converged to make AVs viable—from better sensors to fully electronic driving systems to high definition maps—the necessary solutions will require humanlike perception. A human driver can easily tell the difference between a human about to cross the road versus a bus stop poster with a picture of a person on it. No simple electronic infrastructure will alert you to a dog running into the street versus a harmless tumbleweed. What AVs have always lacked were the perception skills of any child.

Without a robust AI able to classify objects and reliably estimate what those objects will do ("Is that man about to cross the road, or is he just standing there?"), AVs are dangerous murder machines, absently careening down the road, mostly blind to the world around them. In the past, attempts to make AVs absent of AI have required massive infrastructure investments, such as sensors to communicate what speed a car should be moving, when it should stop, and lane markers to tell the car when it's in the lane. And since these cars can't see humans, pedestrians are restricted from entering their driving space. Former attempts effectively envisioned AVs as tiny, single-car trains, or trackless streetcars, but even if we had made such investments, AVs would never drive in your neighborhood. They would never get you home. They were a partial solution at best.

AUTONOMY

Defining what autonomy is can be tricky. Can the sensor-based infrastructure approach even be considered autonomous? What about parking assist? Or lane assist, which allows drivers to take their hands briefly off the wheel and pedals while on an interstate? So far, we have not been specific about what it means for a vehicle to be "autonomous." When we talk about AVs, we usually mean a jump directly to full autonomy. According to the Society of Automotive Engineering (*SAE*), this jump would be level 5 on the autonomous scale, which is the highest level.

SAE Automation Levels

NO AUTOMATION	DRIVER ASSISTANCE	PARTIAL AUTOMATION	CONDITIONAL AUTOMATION	HIGH AUTOMATION	FULL AUTOMATION
Zero autonomy, the driver performs all driving tasks.	Vehicle is controlled by the driver, but some driving assist features may be included in the vehicle design.	Vehicle has combined automated functions, like acceleration and steering, but the driver must remain engaged with the driving task and monitor the environment at all times.	Driver is a necessity, but is not required to monitor the environment. The driver must be ready to take control of the vehicle at all times with notice.	The vehicle is capable of performing all driving functions under certain conditions. The driver may have the option to control the vehicle.	The vehicle is capable of performing all driving functions under all conditions. The driver may have the option to control the vehicle.

Graphic courtesy of SAE International.

As of 2020, level 5 automation is still a few years away, but major initiatives are underway in its pursuit. One side promotes the idea that full autonomy can be accomplished by walking up the ladder of automation. Starting with partial automation (level 2), moving up to conditional automation (level 3), slowly improving to level 4, and eventually reaching level 5. Tesla exemplifies this approach, yet there are dangers to this method. There is a fair bit of research showing that when humans believe they're being assisted, they pay less attention to the task at hand, thus increasing the likelihood of error. This alone should dissuade the idea of consumer-wide automation, even at level 3.

Most other AV companies, led by Waymo and Cruise, are instead making the attempt to jump directly to level 5 autonomy. As a stopgap, AV companies are driving level 4 automobiles on the road under the supervision of trained professional drivers who take over when they encounter new situations, such as inclement weather, and restrict themselves to specific geofenced locations. A variant of this approach is remote monitoring, with humans ready to disengage autopilot when necessary.

Waymo has been running a fleet of level 4 AVs in Phoenix, Arizona, for paying customers since winter of 2018. The cars are called to a pickup location, as in any ride-sharing app, and the vehicle drives up automatically. The user enters and the car takes them to a destination. The car contains a safety technician to take over in anomalous edge cases called *disengagements*. Technically, the same dangers that exist with level 3 (due to human inattentiveness) exist for level 4, just less frequently. It's because of this that persistent human attention is required as a car moves from level 4 to 5, at least until the incident rates drop to acceptable levels. What's an acceptable level? This is an open question, but we can create some reasonable goals.

The average driver is involved in an accident once every 180,000 miles. We want AVs to be better, so if we chose a benchmark of 500,000 miles for an AI driver, it's simply a matter of driving cars around with human technicians involved until a fleet averages half a million miles per disengagement. At that time, we can consider the AV to be level 5 for all intents and purposes—fit for consumer use and safer than human drivers. This alone should cut casualties in half, and improvements in technology can be systemic, jumping from 2× safer, to 4×, 8×, 16×, and so on. The escape velocity of safety will eventually reach what no human driver could match, thus crowning AVs with safety supremacy.

AV Technologies

Modern AVs are a bubbling hodgepodge of technologies and disciplines. Breaking down all the tasks that human drivers accomplish as pilots could fill volumes. Even more so could the technology required to automate those feats. Here's a high-level short list of what human drivers do automatically or with a little training: navigate to a destination, control the car's speed and direction, adjust their driving based on feelings like tire skid, visually perceive the world around them, and predict the likely movement of other agents and react accordingly.

For an AV, these inputs and outputs must be controlled by an array of various sensors and controllers that feed data into a set of filters and artificial intelligence decision makers. Fundamentally, all of these systems exist to help answer in real time the pressing question: "How fast do I go and in which direction?" The answer can be broken down.

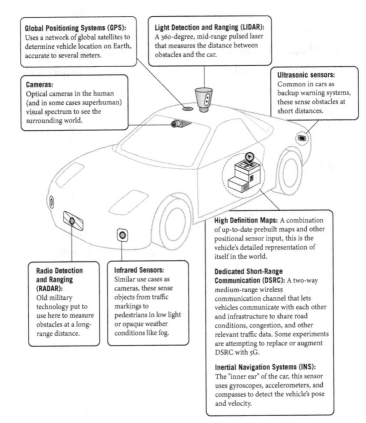

Global Positioning Systems (GPS):
Uses a network of global satellites to determine vehicle location on Earth, accurate to several meters.

Light Detection and Ranging (LIDAR):
A 360-degree, mid-range pulsed laser that measures the distance between obstacles and the car.

Ultrasonic sensors:
Common in cars as backup warning systems, these sense obstacles at short distances.

Cameras:
Optical cameras in the human (and in some cases superhuman) visual spectrum to see the surrounding world.

High Definition Maps: A combination of up-to-date prebuilt maps and other positional sensor input, this is the vehicle's detailed representation of itself in the world.

Radio Detection and Ranging (RADAR):
Old military technology put to use here to measure obstacles at a long-range distance.

Infrared Sensors:
Similar use cases as cameras, these sense objects from traffic markings to pedestrians in low light or opaque weather conditions like fog.

Dedicated Short-Range Communication (DSRC): A two-way medium-range wireless communication channel that lets vehicles communicate with each other and infrastructure to share road conditions, congestion, and other relevant traffic data. Some experiments are attempting to replace or augment DSRC with 5G.

Inertial Navigation Systems (INS):
The "inner ear" of the car, this sensor uses gyroscopes, accelerometers, and compasses to detect the vehicle's pose and velocity.

It takes a lot of parts to replace a human driver.

Control Systems

First, we must consider how a computer can control a car at all. All of the perceptual technology in the world doesn't matter if a car can't move. Initial attempts at AV, such as vehicles in the first DARPA Grand Challenge, required electromechanical actuators to press the accelerator, brake pedals, and steering wheel. They were basically stand-ins for human driver feet and hands. As modern cars have grown increasingly computerized, modern AVs connect directly into drive systems with software that manage a *control system*. One example control system is the

PID (proportional-integral-derivative), which acts as a continuous feedback loop to smooth out simple functions, like cruise control. These are in addition to the necessary *vehicle telematics* for monitoring basic details, like whether or not the battery needs to be changed or checking if windshield wiper fluid is low.

Navigation

One of the fundamental needs in driving a car is *localization*—identifying a car's location on Earth, where it is on the road, what direction it is facing, its bearings, velocity, and more. The impairment of *localization* is what makes a drunk person such a terrible driver. Localization is achieved by crossing the physical *position*, *orientation*, and *velocity* of the car with a high-definition map of the world (containing data such as roads, speeds, intersections, and stoplights).

High-definition maps begin with detailed satellite images. From there, the maps are augmented with more detailed aircraft imagery (think B52s). Next, 360-degree details are added by way of vehicles that drive around the world and map every street, like with Google Maps cars. These maps are then populated with up-to-date public and private traffic and construction information, alongside real-time app data from millions of drivers' smartphones, from apps such as *Waze*. These maps are kept live continually by all AVs in a fleet sharing streams of sensor data.

There's a useful set of older technologies on board that aid in navigation, broadly called dedicated short-range communications (DSRC). This is how autos can communicate with each other vehicle to vehicle (V2V) or vehicle to infrastructure (V2I). DSRC is the technology originally envisioned from the Futurama days, but is still an area of active research in various departments of transportation. DSRC has taken on a new urgency for communicating to increasingly computerized vehicles at scale. Localization and high-definition maps all come together to create a digital model of the world and the car's place in it, called the occupancy grid. The next step is teaching a vehicle to autonomously navigate the world.

Inertial Navigation System

Inertial navigation systems are built with much of the same basic technology that exists in any smartphone. The combination of technologies like odometer, accelerometer, gyroscope, compass, and global positioning technology are available for just a few dollars at any electronic shop. These technologies track the position, orientation, and velocity of the vehicle. Think of an IMU like a human's inner ear, keeping the car oriented and balanced.

The most complex of these are the *global navigation and satellite system* (GNSS), mostly because it involves satellites. GNSS can pinpoint the car's geospatial position on the planet. The most famous example is the global positioning system (GPS) created by the US military, but there are other governmental and commercial geospatial systems that can be used, such as BeiDou and GLONASS. An AV can be designed to use any or all of these systems, so we just call it GNSS.

The IMU, navigation, and control systems all converge to plan and execute the AVs *path* from one location to another. But if we stopped there, our vehicle would be autonomous and highly dangerous. We need to give it an understanding of a dynamic world populated with other objects like vehicles and humans.

Artificial Perception

After control, orientation, and navigation, we need to add artificial perception, which requires various vision systems stitched together in a technique called *sensor fusion*. While there is still some debate about how fancy a vision system needs to be, all AV work assumes that human-level cameras are necessary. Many landmarks on a road assume visual acuity. For a driver to react to a speed limit or stop sign, they first must pass a vision test, and AVs are no different. But some electromagnetic sensors, such as LIDAR (light detection and ranging), provide AVs with superhuman perception of the world around them. These extra-sensory technologies give many industry experts the belief that AVs will eventually best any human driver on Earth.

While humans can only face one direction at a time, LIDAR can detect 360 degrees around the car, through occlusions like fog or dust, and it also works at night. AVs also leverage nonlaser based RADAR for forward collision warning (*FCW*) and avoidance of large objects, but are suboptimal at detecting smaller objects like pedestrians or cyclists. This is the same technology used for parking assist sensors, the kind that beep when you're about to back into a fire hydrant. All of this sensor data is fed into a sensor fusion system (*Extended Kaplan Filters*) with increasing levels of recognition—from signals to characters to symbols—with each level providing a more holistic understanding of the world. These sensors fuse to create a seamless whole: an ever-shifting picture of the world around the vehicle, just like human brains do, creating a conscious narrative from glimpses of perception.

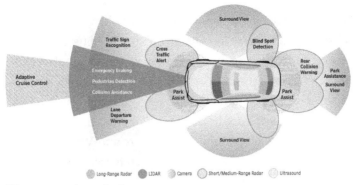

Long-Range Radar   LIDAR   Camera   Short/Medium-Range Radar   Ultrasound

AVs can see so much more than humans can.

More and More and More

Part of the technical complexity of building autonomous vehicles is training an AI to operate confidently and correctly within a cone of uncertainty. This goes beyond alignment of sensor data to move on a high-definition map. The AI must also sense what other objects are doing and predict their most likely path, such as "cars will likely stay on the road, cyclists will be off to the side, and pedestrians will be on the sidewalk." Of course, we know that each of these rules can change at any moment, so the AV also has to detect alternative possibilities.

What may make sense to humans can easily confuse AI.

Then there are the subtle signals and unspoken rules of driving. Cars need to signal to humans where they're going. Real people expect to see human drivers slowly creep into an intersection as a signal to others that believe it's their turn to go. Humans often wave cars on, and sometimes a traffic light is out, requiring a police officer to direct traffic. For all of these conditions and thousands more, we need to ensure an AV is trained and able to respond before we can confidently call it a level 5 fully autonomous vehicle.

NOT THERE YET

We have a long way to go on various fronts before level 5 AVs can reach their true potential. We need to finalize technology needs and legal implications, and must work to find solutions for some dire predictions, like what do we do when millions of professional drivers are suddenly out of jobs? While AVs are one of the most disruptive technologies to appear in the history of transportation, it's the disruptive iceberg below the waterline that will account for the biggest impacts.

EDUCATING AI

There's a bad joke in computer science called the 90/90 rule: the first 90 percent of the work accounts for the first 90 percent of the time, while the last 10 percent of the work accounts for the remaining 90 percent of the time. Autonomous vehicles are called "AI's greatest problem" because the results of minor mistakes can be so dire. Few artificial intelligence problems require such speed and precision with lives directly on the line, and where a model that is incorrect less than 1 percent of the time is still involved in accidents on a daily basis. Training an AI model from 99 percent accuracy to 99.5 percent accuracy could take 10× more effort, while going from 99.5 percent to 99.6 percent can take 100× more, and so on. This is the fundamental reason why, in 2020, AVs seems to have stalled at Level 4 autonomy. Improving the AI to drive safely for 500,000 miles between crashes is magnitudes more difficult than safe driving for just 100,000 miles. The longer a vehicle stays on the road, the greater the odds of encountering rare edge cases, and AI must be trained for most of them. Here are a few scenarios.

Weather

First, there's the issue of weather. While AVs work well in arid climates, even slightly inclement weather drastically increases difficulty. There are simple weathers like rain, and complex phenomena like snow and ice. What should an AV do when confronting a tornado in Oklahoma? We'll never reach a point where the AI powering an AV makes the perfect choice in every scenario, and it's not necessarily a bad thing to prefer some human control in the decision-making process. In extreme cases, a human can weigh options. For example, should they drive through high water or wait for a flood to get worse?

Moravec's Paradox

Problems that are relatively easy for a one-year-old to decide but difficult for a computer are captured in something called Moravec's paradox. AVs need to anticipate the behaviors of others. Is that runner about to cross the road, or are they just stretching at the crosswalk? Is that a police officer standing in the road? If so, are those hand signals directing traffic? Does the added context of a broken traffic light matter?

This lack of human-level common sense expresses itself in many ways. In 2019, it was exemplified with AVs being unable to recognize stopped cars or struggling to make left turns. When an Aptiv car confronts a line of unexpected parked cars in the street, it may also stop because it's under the assumption they are all in line to make a right turn. As for making left turns, these are complicated maneuvers that even human drivers have trouble with. The GM-backed AV company Cruise has made pretty good progress here by focusing on tackling this particular complexity, but it's going to require continued herculean efforts for any new player in the AV space. Or a new cottage industry will need to provide ready-to-use, pretrained AI models.

Security

Security is always a concern with any computer system, but for a computer network that controls a fleet of one-ton terrestrial projectiles carrying people and property, the need for robust security is greater. Think military-grade security with medical-grade testing. There are a million horrific scenarios for criminal access to a vehicle, from theft of shipments via redirecting movement to kidnapping to remote assassinations of ranking officials by causing car accidents. Transporting drugs and victims also becomes less risky, as human agents no longer need to be involved. We need to prevent these real possibilities. We'll also need a robust infrastructure where details concerning an accident are accessible by the proper authorities.

Few people in the world are able to do the work of building AVs, and this has spawned a cottage industry to try and get up to speed. Udacity is an online training course educating engineers on the basics of necessary tools, from creating and leveraging artificial intelligence models to improving the millions of lines of code necessary to create a basic AV. These efforts are important and necessary, and opportunities on the fractal periphery of the AV field will continue to multiply.

EMPLOYMENT

What do we do as a society when AVs automate more and more driving jobs? With over three million people employed in industrial transportation in the US alone, entire logistics operations are likely to be automated, causing mass unemployment in those fields. AVs will certainly create new jobs and industries we aren't aware of as yet, but in the interim, many people will find their skills valueless in the labor market. As a society, we need to have a plan as to what to do when the skills millions of people spent a lifetime perfecting are no longer in demand. As New York City Traffic Commissioner Samuel Schwartz once said, "Cars first, people second, is a mindset that has been difficult to change." Now is the time to change that mindset.

Another industry impacted by AVs is automobile manufacturing. If the promise of ever-present, on-demand AV fleets comes to pass, car and truck ownership will decrease. While a reduction of cars is good news for the environment and for consumers, automakers will make fewer sales, necessitating drops in assembly, parts manufacturing, materials, and all related supply chains.

Drive through any small American town, and you're likely to see a few things. Gas stations, auto mechanics, muffler shops, oil change stations. As these are being consolidated due to shrinking margins and new technology such as electric vehicles, many businesses will have few employment opportunities left. In addition to the lost revenues of small businesses and their associated taxes, municipalities that rely on traffic and parking fees to balance their operational budgets will lose those benefits. This sudden loss of cash will have all manner of effects, from how many police officers the city can employ to cuts in parks and recreation.

Like all emerging technologies, perils must be balanced with promises. Autonomous vehicles will reduce the number of cars and trucks on the road, making travel safer. But the system can also be gamed: empty "zombie" cars might circle the block to avoid parking fees, reducing income for many cities and towns, and increasing pollution in the process. In the US alone, 7.4 million American jobs are tied to the truck driving industry, accounting for around 5 percent of total labor. "Truck driver" is the top job in twenty-nine states, and along with that job, millions of support jobs are also at risk. For instance, 1.7 million automated drivers won't eat at a truck stop pancake house. Throw in warehouse drivers, chauffeurs, train and boat operators, and AVs will have a marked impact on one of the last well-paying jobs that can be done with only a high school education.

ACCIDENTS AND DEATH

AVs' greatest opportunity is reducing the 1.3 million annual casualties due to automobile incidents. While a reduction in traffic deaths is evidently good, there are still downsides. The majority of organs available for transplant come from healthy people who die in automobile accidents. If the traffic death toll drops, the death toll for people on the organ transplant list will increase.

Or consider a case where an AV driving on an icy patch of bridge or another force majeure has a pedestrian in its path. The car can choose to either kill the pedestrian or run the car off the bridge, killing the passenger. How would a computer handle this situation? How would a human? Humans would react in different ways, depending on who the pedestrian was. We might be more willing to strike a single adult but would choose to run the car off the bridge to avoid striking an adult who was carrying a young child. It's unsettling to think about the many permutations of this problem. While you may be unwilling to hit a family, what if your own children were in the car? The thought experiments on human willingness to trade lives is a long-standing one in ethics call the *Trolley Problem*. While its primary historical use case was for creeping out freshman philosophy students, these thought experiments now take on a new urgency. While humans are uneasy about ranking how human lives should be valued on paper, we'll have to do so if we expect a computer to make similar snap judgments, preferably in a way that's transparent to everyone.

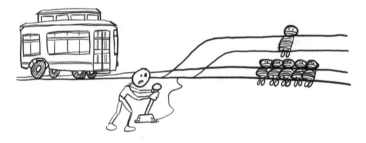

The Trolley Problem.

While AVs have plenty of upside, we can't allow ourselves to be blinded by the promise and ignore the associated perils. Autonomous vehicles must not, as journalist Alexander Kabakov said over the fall of the Soviet Union, "suffer a victory." It's early enough that we can choose to exacerbate the benefits while mitigating the drawbacks. The good news is each downside is another opportunity for us to act on this future and build the best possible outcome. Like any new technology, AVs are not inherently good or bad, but an opportunity to change the world. Let's work to bend that change toward justice for the betterment of more humans.

INCOMPLETENESS

Despite secret wishes of luddites everywhere, AVs are going to happen at scale, and soon in some capacity. The economic upsides are just too great. However, it's possible that the utility we see for the foreseeable future is limited to a handful of corner cases long before we reach the holy grail of self-driving cars capable of driving a child to soccer practice safely. Sertac Karaman, autonomy researcher at MIT once told me, "We may wake up fifty years in the future to find a world where Level 5 AVs never happened." That said, there are plenty of less autonomous transportation options that can still majorly disrupt our economy.

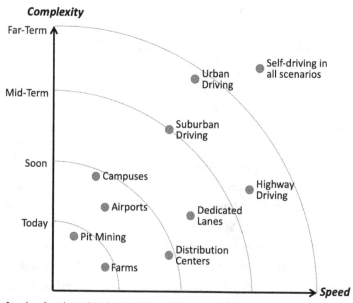

Speed vs. Complexity drive the time horizons of AVs. Compliments of Joshua Siegel.

Long-haul shipping, geofenced transportation like airports, and remotely operated consumer fleets are already happening. We know that with enough infrastructure, we can build cars capable of driving unattended in dedicated high-speed lanes. But it's going to take considerably more effort to build fully autonomous vehicles capable of navigating in all traffic conditions. And until that happens, much of the AV industry will continue to invest billions of dollars on faith. Count me among one of the faithful.

DENOUEMENT

Imagine a world where cars arrived on demand and ownership was nonessential. Those who are stuck in cities with underfunded public transportation could take advantage of personal, direct transportation for a fraction of current fares at 25 cents per mile, opening up more job and education opportunities. Requiring ownership of an expensive and rapidly depreciating transportation asset just to take part in the economy is a terrible injustice.

We can expect city and suburban real estate to transform, as where someone works and lives will no longer be tightly connected. The University of Waterloo in Canada found a direct link between commute time and overall life satisfaction—and no wonder. Who likes commuting? Furthermore, as population increases and land does not, a sizable amount of real estate wasted on parking spaces can be reclaimed for humans. The numerous lots and garages in city cores can be transformed into parks and housing, improving quality of life and decreasing cost of living. And outside of cities, how many homes sacrifice perfectly good square footage for a garage? Personally, I'd much rather have a home gym than a cushy room for my car. The good news is that millennials and Gen Z care more about convenience than ownership, so the mere existence of ride sharing is removing car ownership as a requirement for membership in middle-class America.

Beyond the creature comforts of daily life, the cost of transporting goods will also diminish. This will open up new supply chain potentials. Low-cost, last-mile transportation and autonomous robotic vehicles in warehouses can unlock much of the promise of Industry 4.0. For example, just-in-time custom manufacturing can be powered by more rapid, small-batch shipments. The speed of production will go up, the volume of inventory will go down, and the cost of goods transported will fall. Couple this with personal and business conveniences driven by personal logistics and autonomous vehicles, and the products we buy will arrive faster, fresher, and cheaper.

With the changes happening in autonomy and ride-share fleets, we're finally able to step back and ask ourselves if we ever really wanted cars and trucks at all or just the freedom of movement and the ability to move objects. Safe and available transportation is a cornerstone of liberty in the twenty-first century and beyond. Our descendants will wonder how we were able to survive this bronze age of human-powered transportation and what took us so long to remove our simian brains from the loop. For those of us who have lost friends and family members in automobile accidents, this deep tech cannot arrive soon enough. If we humans can conquer the world on the backs of a handful of well-trained quadrupeds, we can certainly tame our own machines.

FURTHER READING

- *Autonomy: The Quest to Build the Driverless Car, and How It Will Reshape Our World.* Lawrence D. Burns
- *No One at the Wheel: Driverless Cars and the Road of the Future.* Samuel I. Schwartz
- *Driverless: Intelligent Cars and the Road Ahead.* Hod Lipson and Melba Kurman
- *The Driver in the Driverless Car: How Our Technology Choices Will Create the Future.* Vivek Wadhwa and Alex Salkever

7

BEYOND 3D PRINTING

Jelly on the belly, a wand aiming for just the right angle, whirrs and beeps of arcane equipment that only make sense to trained medical professionals. It was an ultrasound like any other. Tatiana Guerra's doctor described what he saw: the forming features of her unborn baby, Murilo—a small potato nose and other cherubic features. Ultrasounds are a rite of passage for modern parents, the first glimpse of their future child, no matter how incomplete that picture may be. Tatiana enjoyed the description. However, being blind, she was unable to see the images of her own baby growing inside. Fifteen minutes into the procedure, a nurse entered the room holding a wrapped bundle, a surprise for Tatiana, who felt around the fabric, smiled, and cried. It was a 3D print of her baby Murilo's face and a plate reading "I am your son" in Portuguese braille. Tatiana was among one of the first blind mothers in history to sense her baby's face before his birth. 3D printing brought her closer to her unborn son, forever changing the world for blind mothers. Today, 3D printing the unborn is a common business. This procedure has recently become popular enough that 3D printing capabilities are built into some ultrasound machines, like the *GE Voluson E10*, and healthcare providers can easily print a 3D model of a fetus based on volumetric scans of a child in utero.

Making tools is how we humans have measured our progress from the Stone Age to the Bronze Age to the Iron Age. If aliens landed on Earth, they too would likely categorize our species as "the one that makes the tools." There are four known ways to make physical things: shape materials (molding, extruding), subtract materials (carving, milling), add materials (assembly, 3D printing), and self-organize (growing things). Most things we use have been built using one or more of these processes. But we're witnessing the birth of a new form of tool making—additive manufacturing (*AM*), which is by far the newest. Since assembly is technically also additive manufacturing, we'll use the terms *3D printing* and *additive manufacturing* interchangeably in this chapter.

3D printing works by fusing materials layer by layer, controlled by a computer, for the purpose of building complete and novel shapes with very little waste. Like the rings of a tree, this process builds up thin layer by layer over time, from a simple seed to a complex structure. It's a quick, inexpensive, and durable method for printing novel 3D shapes.

In general, it's expensive to ship tools into space, and will continue to be expensive for quite some time. Moreover, it can take months or years to get tools to where they need to be. At such high costs and time, the ability to create new tools on demand is a game-changer for the space industry. In 2016, a company called Made in Space, in association with the International Space Station (ISS), deployed a low-gravity additive manufacturing facility (AMF) on the ISS. The AMF has printed hundreds of minor tools, replacing broken wrenches and screwdrivers and, in one instance, printed an emergency splint in 2017 to treat an astronaut's broken finger, marking the first medical supply built in space. If humans ever dream of visiting other planets like Mars, this capability can't be optional. But printing in space is not the only harsh climate where 3D printing has made an impact.

In 2013, film producer and inventor Mick Ebeling boarded a plane, nervously stowing boxes of components away on a flight. Against United State advisories, he traveled on a roundabout trip into the middle of war-torn South Sudan, hopping planes and trucks for hours to the Nuba Mountains where he set up shop. Mick was touched by the horrors of the war's high civilian collateral and stories of children who lost limbs. He was especially concerned with one boy, a double-amputee named Daniel. Backed by his organization, *Not Impossible Labs*, he designed and set up a 3D printing laboratory to print artificial limbs for amputees. It's difficult and expensive to design and ship prosthetic limbs to war zones, so Mick taught locals how to operate the machines and use open source prosthetic models to help the injured with artificial mobility. Now, with the help of his prosthetic limbs, Daniel works in the laboratory to help design and fit limbs for more than 50,000 amputees in the area.

We could go on and on. Yes, these are emotionally charged stories, and they are a small example of the important use cases that are difficult or impossible without 3D printing. From inner space to outer space, from war to peace, 3D printing is revolutionizing how we make objects, democratizing the technology and allowing anyone to shape available materials in a cheap, efficient, and standard way, without the need for scale.

BENEFITS

It's hard to overstate the rapid evolution of 3D printing. The first numerical control machines were built around the 1940s, adding electronic logic units to create computer numerical controls (CNC) around the 1960s. These machines allowed for mass production of increasingly precise parts in subtractive manufacturing, such as cutting, boring, and milling. In the late 1980s, the first examples of what we now call 3D printing burst onto the scene, using computer controls similar to some CNC. In fact, some of the files used today, like gcode, are similar between both CNC and AM. The purpose of these early attempts was rapid prototyping, creating something called a stereo lithography apparatus (SLA) and, later in 1989, fused deposition modeling (FDM). Throughout the nineties, several improvements arose, but these were mainly focused on large industrial uses, and these devices cost tens or hundreds of thousands of dollars. Then, in the first decade of the 2000s, something changed. The rise of internet communities and the open source ethos championed by the software community started infecting other domains. The group of artists and hobbyists calling themselves Makers began openly sharing and improving upon a fused deposition modeling machine for personal use. It was little more than a hot glue gun affixed to three rails capable of moving the nozzle to any point in a small volume. Then a small computer controlled exactly where the nozzle deposited a small droplet of plastic onto a surface. Once a layer was put down, it moved up a fraction of a millimeter and printed out the next layer, repeating until an entire shape appeared.

Despite these early designs being openly available, they still required a wide cross section of capabilities, like working knowledge of electronics, basic engineering, a bit of fluid dynamics, and programming. That and lots of trial and error. In 2009, two simultaneous events helped turn this niche hobby into the seeds of a real industry. First, crowdfunding, namely Kickstarter. The year Kickstarter launched, several 3D printer designs became available for the public to buy, both as DIY kits or fully built printers. Second, and equally important, this was also the year that the original 1989 FDM patent expired. Seemingly overnight, a cottage industry arose that sold printers, materials, and predesigned 3D models.

The industrial revolution was fueled and driven by standardization of parts. Ford couldn't have created an assembly line of millions of Model T cars unless every bolt and nut—and by extension, every wrench—came standard. Before the assembly line, most products were hand built individually by artisans, adhering only to the most general standards of practice. Thanks to the internet, it became easy to share 3D models. There were no specific industrial standards; standardization became an emergent property based on what worked for the most people. Standards are increasingly aligning for file types, printer sizes, and materials, all absent of central authority.

THE SMALLEST BATCHES

Like many teenagers in the rust belt of Indiana, the summer job of choice was in auto part manufacturing. Despite the heat and repetitive nature of the work, it paid better than a gig at Burger King. My particular department stamped out car door frames. All day, every day in the broiling heat, I lifted out a sheet of steel, pressed the stamp, and removed the newly shaped part. Eight hours a day every day, the only diversions occurred when defects arose. We stopped the press, and engineers adjusted or replaced the dies. Replacement parts cost up to $50,000 each, close to the yearly wage of my coworkers.

Running large batches of similar parts has been the status quo since the late 1800s. While this is cost effective for a large factory making millions of identical parts, it doesn't work when you need a small run of a few parts, or you need to make a high variety of components. Tooling is expensive. Dies, casts, milling bits, and the like are all expensive to purchase, maintain, and exchange at a moment's notice. If you wanted to build your own projects, you were beholden to whatever happened to exist in bulk or pay a premium for an expensive handmade part—3D printing has opened the world to inexpensive, custom, low-volume manufacturing.

The first use cases of 3D printing were rapid prototyping. The machines were expensive, but for certain part makers, the ability to experiment or show customers detailed concepts was worth the cost. But these were always the first steps to the end goal of an eventual mass-production run. Today, you can make custom parts in 3D without having to make more than one, which is very attractive to the Maker movement.

Marketplaces like CGTrader allow suppliers to sell 3D computer-aided design (*CAD*) models for professionals to manufacture in any quantity. For hobbyists, Thingiverse lets users trade and modify 3D models for free. The 3D model ecosystem is an important component of the 3D printing revolution, allowing people to start quickly and cheaply without any 3D CAD expertise. Of course, copyrights must still be respected, as the now defunct PrintABrick website learned when they attempted to sell 3D prints of LEGO bricks.

While the cheapest 3D printers cost about $200 in 2020, high-quality prints are still available in small batches. Thanks to the easy availability of 3D digital models, several boutique 3D printing services have arisen over the years. Shapeways is additive manufacturing as a service. Users provide the 3D model they want, and Shapeways will build a professional-quality object made from a wide selection of materials, from plastics to metals to carbon. It seems that no niche is too small for 3D printing, as FigurePrints is a growing business that deals in custom characters for the video game *World of Warcraft*. Players pay a premium for custom prints of their own avatars, wearable equipment they won in a game, and mounts on a stand featuring their character's name.

APPAREL

Even before the Luddites smashed loom machines in the early 1800s, fabric has been at the forefront of many technology revolutions. That spirit of innovation thrives in the on-demand and custom knit space. So much on-demand knit clothing via knitting machines has been tried by companies as large as Adidas and as small as Ministry of Supply and Kniterate, that it's easy to forget custom knits are already an industrialized process. Nike's Flyknit shoes are made by machines that run on code. Every shoe, in theory, could be customized. But there are more types of additive manufacturing in the apparel space than just automated knitting.

Hollywood is getting into the act. The film *Black Panther* featured a character, Queen Ramonda, whose outfits were made with a technique called Selective Laser Sintering (SLS), where lasers form the product within a layer of nylon powder. Another technique is being leveraged by a company called Native Shoes, who are working on 3D printed shoes made of recycled material alongside MIT's Self-Assembly Lab. These shoes are printed within a liquid gel bath. Both of these techniques are interesting in that the substrates act as support materials, allowing for intricate and detailed shapes, as well as supporting the kind of flexible, comfortable material you'd want for apparel.

For something more home brewed, Danit Peleg, a designer based in Tel Aviv, has designed a line of clothing that can be printed by a more common 3D printer. While her designs may be a few years away from popular use, some Makers are already printing custom T-shirt designs that literally pop by placing fabric directly onto a 3D printer and building a plastic design. And for the environmentally conscious, not only can 3D apparel be made with recycled materials, but 3D printing can also stitch seams and print replacement components like zippers and buttons, extending the life of your favorite outfit.

MEDICAL

When the borders of Italy were locked down in the midst of the coronavirus pandemic, the need for medical equipment skyrocketed. Due to disrupted supply chains, a hospital in Northern Italy could not get a valve required to repair a broken resuscitation machine. After a few phone calls, AM company founder Cristian Fracassi carried a 3D printer to the hospital and cloned and printed the missing part within a few hours.

Medical 3D printers can print more than plastic pieces. In 2015, the FDA approved a 3D printed, rapidly disintegrating epilepsy pill called Spritam. Created by Aprecia Pharmaceuticals, it was a practical example of how 3D printing can take part in the field of personalized medicine. Later experiments have focused on custom shapes that make pills easier to ingest for kids or for those with a limited ability to swallow.

More exciting breakthroughs have emerged around the 3D printing of skin, bones, and organs. Rensselaer Polytechnic Institute and Yale School of Medicine partnered to create a printed skin made of living cells with correct layers, later with vasculature, that can be grafted into humans while minimizing organ rejection. NYU School of Medicine printed a ceramic scaffolding for bone to cling to, promoting correct growth of the patient's natural bone structure. Previously impossible bone transplants have succeeded as prints, such as new skull parts and replacement vertebrae. These bones can also grow, an important trait for bone printing, especially in children.

Other 3D printed parts are real as well. Newcastle University scientists successfully printed a human cornea, something that ten million people worldwide need to regain their sight. Pushing the boundaries even further in a 3D printing story that borders on science fiction, researchers at Princeton 3D printed a bionic ear that merged tissue and electronics, allowing a person with this transplant to potentially hear ultrasonic frequencies once reserved only for dogs.

Bioprinting was first attempted in 2003, but since then, it has expanded in scope and use cases—livers and kidneys have been 3D printed, as well as other organs, with a variety of techniques in various degrees of study. This is an important and necessary area of research and is already attracting startups like Organovo. As a practical matter, organ printing cannot arrive soon enough. As we explored in Chapter 6, it's estimated that autonomous vehicles will save millions of lives, which is a good thing, but has the negative impact of shrinking organs available for donation—3D organ printing is humanity's hedge against this eventuality.

ARCHITECTURE AND CONSTRUCTION

Picture a world where you can design your own home on a computer. You pick the size, the rooms, the shape—straight edges optional. Then imagine that once you've designed it (and walked through it virtually in VR), it could be built for as little as $6,000 in forty-eight hours. Welcome to 3D printed buildings. This example is not science fiction, by the way. A 1,900-square-foot home matching the specs in the image was printed in early 2020 by a New York–based company *SQ4D*, using a technology called Autonomous Robotic Construction System (*ARCS*). Like standard 3D printing, it builds up the walls one layer at a time with a quick-dry concrete mix—complete with spaces for doors, windows, and insulation.

SQ4D's 3D printed home. Photo via SQ4D.

The speed and reduction in cost is partially due to the construction material, a concrete composite that's fire-resistant and flexible enough to withstand extreme weather (even seismic activity). It also requires considerably less labor—it takes fewer people to construct a house with this technology. Of course, it still requires experts to install components like siding, wiring, and plumbing…for now. Never to be outdone, a larger, 6,900-square-foot, two-story office space was printed in

Dubai, requiring only three workers to build it. This is a curvy, structural marvel, breaking the constraints of square walls with ease. Dubai plans to ensure that 25 percent of all buildings are 3D printed by 2030, placing it firmly in the leader category in this space.

The history of technology has been about making luxuries more accessible to the masses, and few things are as luxurious as having a nice place to call your own. With this in mind, additive manufactured homes are helping to fight poverty. In Tabasco, Mexico, developers are printing an entire neighborhood of fifty houses for impoverished families. It's a collaboration of charities called New Story and ÉCHALE, with the help of 3D printing company ICON. The Vulcan II printer can construct two houses at once with just twenty-four hours of work, which is considerably faster and sturdier than hand-built alternatives and for a fraction of the cost.

Beyond the structure itself, home goods from furniture to sconces are also being printed. I have a 3D printed lamp on my desk, which gives off a unique spirographic light. With such a wide array of utility, speed, cost-effectiveness, and reduction of waste, 3D printing has a solid future in the construction business.

FOOD

"Tea, Earl Grey, hot!"

—CAPTAIN JEAN-LUC PICARD, *STAR TREK: THE NEXT GENERATION*

Like many emerging technologies from smart tablets to AI-based voice interfaces, automated food generation has roots in the science fiction show *Star Trek*. A device called a *replicator* could create any dish the user requested. So it makes a bit of sense that some of the original research for printing 3D food started at NASA in 2006, with the intention of feeding astronauts a variety of foods based on a common set of ingredients, with very little waste.

Over the years, 3D printed food has become a cottage industry. Food printers can range in styles. There are basic printers, which can use sugar and chocolate (Mmuse) to craft complex shapes, like a plate-sized Palace of Versailles dessert printed in sugar. For the more specialized, there are pizza- and burrito-focused printers and the Norwegian PancakeBot 2.0. For the complex, there is Giuseppe Scionti's Novameat, a prototype printer that uses plant products to print realistic textured steaks.

A 3D printed Versailles Palace in sugar.

As Earth's population grows, 3D printers can also help reduce waste. Artist Elzelinde van Doleweerd used food waste to create new printed foods. These were foods too ugly or poorly textured to sell, but they could be recycled and printed into beautiful foods that some would pay top dollar to eat. In the television show *Upload*, about a near future world where flying drones are omnipresent and food printers are common kitchen appliances, characters discuss a Jamie Oliver printed steak recipe. It's a realistic look at the banality of deep tech in the future, where the revolutionary becomes as commonplace as a refrigerator in modern times. Gourmet food in your home, high in nutrition, low in waste, made from recycled ingredients. Leftovers will never be the same.

INDUSTRIAL AND SUPPLY CHAIN

There's a rule in large-scale distributed computing: it's easier to ship the algorithm to the data than the data to the algorithm. When you're dealing with exabytes of data smeared across servers spanning the globe, it's more efficient to send a small equation to each of the servers and sum up the few results than it is to shove petabytes of data through a single server and run an equation.

The benefits of 3D printing can function in a similar manner. It can be easier to ship a box and raw materials to certain locations for people on the ground to manufacture what they need, rather than to take a set of custom orders, make them in small batches at a specialty maker, and then deal with the logistics of shipping those complete products to the end user. Just in Time (JIT) manufacturing, the reduction of shipping costs, and speed of onshore manufacturing was the breakthrough logic that brought Mick Ebeling to print 3D limbs in Sudan, or why the ISS uses AM in space. In an age where people are starting to demand smaller environmental footprints, higher customization, and faster turnaround, consumers will likely force companies to adopt 3D printing as their predominant method of make.

Let's think a bit farther ahead. Imagine a network of microfactory 3D printers dotting the globe, able to make anything out of any material, like Shapeways. This creates an opening for an entire JIT supply chain: consumers choose what they want to make and it's made on demand rather than made far away or held in a distribution center. The broker, the owner of the design, the microfactory, and the logistics company (why not a self-driving fleet?) are paid for their respective services—each small player makes a living with no central authority. This is the final stage of the Industrial Revolution: full democratization of ideas to market at scale, open to anyone.

But it's not only leaner manufacturing and slimmer supply chains that are driving industrial uses of 3D printing. The first and still most prominent use case for 3D printing is rapid prototyping. The ability to quickly build and test is the true superpower behind the company *SpaceX*. Elon Musk, billionaire, inventor, and true-life Iron Man showed off the SpaceX rapid prototyping facility. The 3D parts are designed and modified using hand gestures and virtual reality, run through simulation testing, and then the metal components are 3D printed on demand. Instant rocket parts. And it's not just Silicon Valley startups in on the action either. The Air Force saved millions of dollars by printing a fifteen-dollar part that could check for fuel leaks.

To recapitulate, apparel, medical, buildings, food, and industrial are but a few of the use cases where 3D printing is making big inroads. We discussed a few others in the introduction, like space travel—3D printing is changing lives and even cultures. Now let's dig into how this deceptively simple technology works.

TECHNOLOGY

While versatile in their ability to support a wide variety of materials and shapes—and thus use cases—3D printers are fundamentally similar. They break down a 3D digital model into layers, then print one layer at a time in 2D (think a standard inkjet printer). Once a layer has been made, it moves up (or sometimes down, depending on the printer style) and 2D prints the next layer. This process repeats until you end up with a 3D object.

Although those are the fundamental steps, the devil is in the details. Printing in 3D is still an industrial process, although heavily automated and shrunk down to the consumer level. Using a 3D printer requires a bit more care than the mass market might be ready for just yet. There are complexities in creating printable 3D digital models, but these are getting easier all the time. With zero CAD experience, I've been able to build quite complex shapes using an app called Shapr3d with a basic iPad and Apple Pencil. CAD design has become so banal recently that there are several websites offering this ability for free, requiring little training and no specialized equipment.

There's no perfect checklist, and 3D printing is still in the early phase of technology where it's much more art and rules of thumb than immutable rules. But there are a few considerations to keep in mind when attempting to land on the right additive manufacturing approach. The first concern should be the necessary characteristics of the fabricated object you want. Mechanical, scale, and cosmetic requirements should be use-case driven. Second, material selection is a driving factor in deciding which process to use. Third, consider the geometry of the shape. This includes level of detail, wall thickness, precision, and the like. Finally, the cost, which beyond materials is also a function of aspects like printer cost, print speed, and cleanup effort.

Printing in 3D is, at its core, computer-controlled deposition—meaning it's a process of depositing materials one layer at a time. There are collections of techniques that fall under this rather broad process that we'll look into now, starting with categorizing the types of printers, followed by a peek at 3D digital modeling.

AM CATEGORIES, COMPLIMENTS OF ISO/ASTM

People love to categorize things, and nowhere will you find a greater love for this than with the International Standards Body (ISO) and the American Society for Testing and Materials (ASTM). As famed statistician George E. P. Box once said, "All models are wrong, but some are useful." ISO/ASTM created a set of additive manufacturing process categories that are useful for thinking across the range of 3D printing. These groupings shouldn't be thought of as hard delineations, but as scaffolding with unclear borders. The categories are material extrusion, vat photopolymerization, powder bed fusion, binder jetting, material jetting, directed energy deposition, and sheet lamination.

Modern devices sometimes cross categories, especially in the higher-end AM market from companies like StrataSys. On the other hand, some AM-like capabilities, like FBR's Hadrian X brick-laying robot, are accounted for in this classification with some imagination. You might call stacking whole bricks to construct buildings "material extrusion." Overall, these categories are useful for understanding the landscape of technical possibilities and helping you further investigate which style of AM is most relevant to your needs.

Material Extrusion

Material extrusion is a process where material is selectively squeezed through a nozzle or orifice, and it's what many people think of when they imagine 3D printing. Picture a hot glue gun where each droplet of glue is precisely controlled by a computer and moved around by a pair of railings. That's the basic structure of some of the first consumer 3D printers like the MakerBot, also called Fused Deposition Modeling (FDM). This is the kind of printer you'd find among hobbyists: the device pushes a plastic-like filament or wire through a heated nozzle, causing it to melt. This basic structure is useful for materials like thermoplastics and soft metal filaments, although some of these printers can push out liquid materials as well.

What's exciting about this style of printer is its versatility and relatively low cost, both to build and manage the printer and the materials. My home 3D printer, an Ender 3, costs around $200 in 2020 and is completely open source. It's even possible to print many of its own parts, in a sort of primitive cloning capability.

Material extrusion printers can be scaled up in size to support concrete blends for building houses or modified to support edible products for printing food. It's no wonder that this sort of printer has captured the public imagination and, more than any other, has triggered the AM revolution.

Vat Photopolymerization

Vat photopolymerization, despite its complex-sounding name, is a simple process in principle. If you've ever had a chipped tooth or a cavity, you may be familiar with photopolymers. The material is a paste formed into the shape of your tooth, and when it's introduced to ultraviolet light, the material hardens, is shaped with a dental bur, and—voilà!—instant tooth repair. Now imagine how this curing process would be helpful in building a 3D printer.

Targeted light pulses aimed at a thin vat of liquid photopolymer resin hardens some of the liquid layer into a complex 2D shape. This process is called light-activated polymerization. A mechanical arm pulls the hardened structure a few nanometers vertically, and the light repeats the process for the next layer. Over time, you build a 3D structure made of solid resin.

Unlike material extrusion, vat photopolymerization does not rely on a layer of thin material deposits from a nozzle, but rather, the printing process controls light through a device called a mirror galvanometer, or *galvo* for short. Furthermore, unlike material extrusion that tends to print from the bottom up, vat photopolymerization tends to print the top layer first and work its way to the bottom.

The oldest kind of additive manufacturing, vat photopolymerization was first developed as stereolithography (SLA) in the 1970s by Dr. Hideo Kodama, who sadly was unable to file a patent due to funding issues. Years later in 1986, Chuck Hull was awarded a patent for SLA and, shortly after, started a 3D printing company. This event is commonly considered to be the birth of modern 3D printing. At the time, the focus was rapid prototyping but has since evolved beyond these early use cases. Now, vat photopolymerization is common for cases where the output requires a smooth surface and fine details, such as with medical applications. The dental maker Invisalign has been using an SLA process to mass produce unique aligners for years.

Powder Bed Fusion

The previous two 3D printing categories focus mainly on types of plastics. But sometimes metal components are necessary, and powder bed fusion is a good solution. Like stereolithography, the 3D object is constructed one layer at a time from a vat of material. But instead of a shape being cured and pulled up from a liquid resin, an energy source fuses a powder to the desired shape one layer at a time, before another layer of powder is laid down. There are a few benefits to this approach. First, since high-energy sources like lasers are used to melt the powder, the variety of materials can extend beyond powdered plastics— powdered metal materials can be included as well. Second, due to the kinds of materials and high melting points, powder bed fusion techniques can create stronger parts, although at a higher cost than other categories. Third, since the 3D pattern being fabricated sits inside a bed of powder, the need for support structures all but disappears, reducing waste. Much of the time, the unused powder bed can be reused.

Selective Laser Sintering (SLS) is the canonical approach to powder bed fusion, and it involves using a directed laser as the energy mechanism to fuse the powder. Like stereolithography, the laser is directed with galvos. Direct Metal Laser Sintering (DMLS), Selective Laser Melting (SLM), and Electron Beam Melting (EBM) are other varieties of powder bed fusion.

Directed Energy Deposition

If material extrusion and powder bed fusion had a baby, you'd end up with directed energy deposition. This process focuses thermal energy to fuse materials by melting powders or wire as they are being deposited. The energy source defines the kind of machine, from laser (Laser Engineered Net Shaping, or LENS) to electron beam (Electron Beam Additive Manufacturing, or EBAM) to plasma arc (Plasma Arc Directed Energy Deposition, or PA-DED).

What's interesting about this process is its ability to repair existing parts rather than simply fabricating new components. A directed energy deposition machine on a shop floor can be used to repair anything from handheld tools to turbine blades. You can, in the simplest terms, think of DED as a type of autowelding machine.

Binder Jetting

Binder jetting as we know it today was developed by MIT in 1993. Then, in 1996, the company *ExOne* expanded the technology to work for printing metals. It's sort of a cross between powder bed fusion and a classic inkjet printer you'd use at home. Binder jetting fabricates a complex 3D geometry without the need of supports. Rather than fusing each layer with a high energy source, it instead uses liquid binding agents to join powdered materials. Think of microscopic droplets of glue dripped onto a layer of baby powder to make a flat shape, then repeated layer after layer. That's the general idea. It's unique from most 3D printing processes in that it doesn't require heat or light, just chemical processes. This helps reduce some of the distortions and stress concerns, like warping, that can arise from processes requiring heat.

Binder jetting can be useful for constructing products from powdery materials like sand or ceramics into massive structures such as a 3D printed pedestrian bridge in Madrid. While it can be useful for fabricating some metal parts outright, it can also print 3D sand casts, used to mold molten materials like iron or steel. This is a modern update to the eons-old foundry process.

Material Jetting

Material jetting has similarities to material extrusion and inkjet printing. Like material extrusion, material jetting deposits droplets of material onto a build plate, where they are cured when exposed to ultraviolet light, hence requiring a photopolymer resin like stereolithography. What makes material jetting distinct from material extrusion or stereolithography is its ability to support different materials fabricated in the same object. Unlike many other forms, material jetting also allows for full-color printing. If you've ever seen a home 3D printer, they tend to work in only one color.

Drop on Demand (DOD) is a type of specialized material jetting process. But rather than general purpose, it's designed with some heads that are involved in dissolvable support material deposition and other heads that fabricate the model. DOD printers are often used for building lost-wax castings.

Sheet Lamination

As the old saying goes, "You can have a job done fast, cheap, or good, but you can't have all three." In the 3D printing world, sheet lamination is a clear case of fast and cheap. The materials of choice in this process are paper and aluminum foil, and the machine uses lasers to cut shapes and glue them together one sheet at a time. The interesting thing about the output of sheet lamination is that, depending on the geometry, enough sheets of paper can start to gain some of the strength and characteristics of wood. And because of the heat and harsh chemical-free process, it's easy to add color to the output through standard inkjet technology.

Sheet lamination is sometimes known as laminated object

manufacturing (LOM). For architectural models and product prototyping, sheet lamination is one of your best bets. Some people have experimented with this technique to build furniture as well.

This may have felt like an exhaustive overview, but 3D printing is such a broad field. It's important to see that the industry is so much more than the desktop 3D printers your nephew plays with. It is the next great frontier of manufacturing. To recapitulate, the ISO/ASTM AM categories are material extrusion, vat photopolymerization, powder bed fusion, binder jetting, material jetting, directed energy deposition, and sheet lamination. There's no quiz, but it's helpful to be familiar with the seven categories at a high level if you're interested in making a play into the AM space.

3D DIGITAL MODELING

The first thing you'll need before printing anything in 3D—from a plastic toy to a house—is a 3D model of the object. Find a digital model you like online, design it yourself, or commission someone else to render your idea. Getting the general shape is a start, but with the maps of old, there be dragons here. When designing the 3D digital model, you need to keep in mind what the finished product will be used for. That will drive the kind of materials necessary, being mindful of postprocessing like grinding and filing. These details, in turn, will dictate particulars like wall thickness or infill.

CAD software, like SolidWorks or Rhino, are highly detailed yet complex tools for designing custom 3D manufacturable models. Unless you absolutely require custom objects, like rocket engine parts, the more common method is heading to a model marketplace where 3D models are bought and sold. There are also several open source models freely shared online, with the most popular site being *Thingiverse*.

Different 3D printers support different file formats, but there are two major kinds: 3D mesh and NURBS (nonuniform rational basis-splines). Three-dimensional mesh files like OBJ or STL are a collection of tiny polygons that create a shape. This is the file type you'd see rendered in video games or animated movies. They look good, but beauty is only skin deep. NURBS are more of an industrial file format. Rather than polygons—which are detailed but ultimately approximations of a surface—NURBS are mathematical functions that represent a true surface. They can also be more difficult to work with. Some printers accept various formats and convert them internally, but for the best results, practitioners often like to control file details, such as defining supports and slicing before printing.

This was just a brief overview of the physical and digital technologies surrounding additive manufacturing by way of 3D printing. It's a tough topic to deconstruct and, like the subject of extended reality, hard to grasp without hands-on experience. Hopefully, you're intrigued enough to try it out for yourself. The benefits of this deep tech will only grow over the next decade and beyond.

ISSUES

With all of the flexibility in materials, quality, speed, and cost of additive manufacturing, it's easy to believe it's a panacea of all future manufacturing techniques. The process has clear values and contains decades of revolutionizing, from industrial manufacturing to construction, from supply chain to food service.

However, these benefits have tradeoffs. Some are temporary (like slow print speeds) while others are probably permanent (like the black market of 3D designs).

Time and materials are the two key indicators of cost in any production process, and those indicators are amplified when attempting to mass produce a product. They are also two of the big weaknesses of 3D printing as it stands today. Beyond a relatively slow fabrication speed and limited and expensive material options, 3D printing also suffers from a general lack of precision and geometric choices, and there are gray areas legally when it comes to printing an easily transferable digital model on demand.

PRINT SPEED

Three-dimensional printing processes are slow. Depending on the materials, geometry, process, and other variables, it can take hours to fabricate even the smallest nut and bolt assembly after you factor in post processing. During the early days of the coronavirus pandemic, I attempted to help a local hospital by 3D printing respirator masks. However, each one took around four hours to complete, and it was orders of magnitudes faster to simply sew or weld parts together.

But not all hope is lost. Joseph DeSimone's TED talk, "What If 3D Printing Was 100x Faster?" filled the auditorium with delight, causing audience members to audibly gasp. While he spoke, Joseph fabricated a plastic ball in under ten minutes. The process he demonstrated was called CLIP, a type of vat polymerization that is faster than other prior processes. But we could still do better. In 2020, the Swiss Federal Institute of Technology Lausanne created a new method that can print small, soft objects in under thirty seconds.

LIMITED MATERIALS, MULTI-MATERIALS

3D printing is a good option if you need plastics or metals, and is a serviceable option if you need paper or ceramic. But iron requires the classic added steps of casting and milling. Glass is in the early days of experimentation but not yet close to the variety the industry needs.

Beyond material limitations, there are also limits to parts that are made of certain materials. You can't easily 3D print a car, for example, since there are steel frame requirements with rubber, aluminum, carbon, and other necessities. At the end of the day, your best available option is to 3D print the components and then go back for robotic assembly. Of course, this would be a very expensive car due to the aforementioned cost and speed issues.

PRECISION AND GEOMETRY

Three-dimensional printers are, on the whole, not very precise for many industrial needs. Forget the nano scales of electronics—many 3D printers have difficulty getting much below the millimeter scale, which is a thousand times less precise than the submicron scales necessary for some aerospace or medical components. However, things are improving. A company called Exaddon printed a prototype millimeter scale copy of Michelangelo's *David*. This has only succeeded with a single material so far—ionized liquid copper—but it's a big, tiny step in the right direction for printing precision.

Exaddon's CERES 3D printed David in a fraction of a millimeter width.

Certain materials and processes also don't lend themselves well to 3D printing. For instance, due to the heat necessary for FDM, there is a minimum wall thickness allowed before the shape distorts, limiting the kinds of geometries modern printing methods allow. There are endless examples, however, of teams looking to break these barriers. Researchers at ETH Zürich are creating complex glass objects with 3D printing. Digital Manufacturing and Design Centre from the Singapore University of Technology and Design are using data to let designers build complex geometric and material structures that exhibit new behaviors hitherto unmakeable. This is an active area of research, and increased precision and new geometric breakthroughs might very well be the first true killer application of 3D printing at scale.

LEGAL GRAY AREAS

The world broke new ground on May 6, 2013, when a group calling themselves Defense Distributed released open source designs for a working, 3D printable gun named the Liberator. The US Department of State tried to force the removal of the digital plans, kickstarting an international race to decide how to proceed in this new world. In 2020, the United States is still undecided about allowing individuals to print guns from home and share the designs, but countries like Australia and the UK have been clear on their illegality.

Another pressing concern is how easy it is to break intellectual property (*IP*) with 3D printing. While specific CAD files are often clearly IP, what about 3D scans and fabrication of existing products for personal use? While digitally copying a book or music is well understood to be piracy, is it theft to print a broken plastic valve for a hospital respirator? There are shades of gray between copyright piracy and fair use. And where shades of gray exist, black markets dwell. And in this black market, data is much easier to move than physical goods. Why get caught selling counterfeit products when you can sell the plans and let consumers build their own? Where do copyrights end and patent rights begin? These and many other questions will be solved by many different countries over time. The more consistent we can align the rules internationally, the sooner they will be solved.

Despite the downsides, speeds will increase, materials will expand, and volumes will rise. Between the customization at scale and the reduction in waste, the effort put forth is worthwhile.

DENOUEMENT

To build tools is to be human. For the first time in tens of thousands of years, we're at the beginning of a new production process called additive manufacturing or 3D printing, which enables decentralized manufacturing. Like open source software or blockchain, decentralization removes choke points, reduces barriers to entry, and increases the variety of options and the robustness of value chains. The tradeoff is reduction of economies of scale (a short-term issue) and variability in quality. This has been made possible by cheap, standard electromechanical parts from servos to sensors, and small, powerful computer components like Arduino. The same technology that's driven the rise of IoT opened the door to mass market 3D printing. In a very real way, the AM revolution has ridden on the backs of smartphones, IoT, and the internet. Beyond anthropological interest, anything capable of affecting the $12 trillion per year manufacturing sector is worth taking notice of.

Additive manufacturing has broad implications. It could be something as singular as helping a blind mother see her child for the first time or as game-changing as helping make extended space travel possible and helping children in war zones lead better lives. It also means that anyone can print untraceable weapons or illegally pirate physical goods with as much ease as copying a digital file. It's an exciting time to be alive—a little bit Wild West. In some ways, we're reverting to an earlier artisanal age, where individuals make the things they need. But what's new and exciting is that this capability can be scaled for nonspecialists, thanks to the internet, and improved by anyone. And with increasing specializations such as Penn State's master's in additive manufacturing, these skills will only improve in the aggregate. It's now easy to imagine a day where every home will come with an easy-to-use fabricator or two for printing objects from the small to complex, from knickknacks and replacement parts to clothing and even food. The supply chains of this future will look very different, delivering raw materials rather than completed projects.

FURTHER READING

- *3D Printing: Modern Technology in a Modern World.* Raymond T. Reeves
- *3D Printing.* John Jordan
- *Designing 3D Printers: Essential Knowledge.* Neil Rosenberg
- *The 3D Printing Handbook: Technologies, Design, and Applications.* Ben Redwood, Filemon Schöffer, and Brian Garret

8

PICKING QUANTUM LOCKS

"Hell, if I could explain it to the average person, it wouldn't have been worth the Nobel Prize."

—RICHARD FEYNMAN

Richard Feynman at Cal Tech, 1986.

"I opened the safes which contained all the secrets to the atomic bomb: the schedules for the production of the plutonium, the purification procedures, how much material is needed, how the bomb works…the entire information that was known at Los Alamos: the whole schmeer!"

—RICHARD FEYNMAN

In addition to his storied career as a physicist and public intellectual, the charismatic polymath Richard Feynman was an accomplished lock pick. His life reads as a series of hacks, finding shortcuts to solve difficult problems that stumped other experts of the day. At the top secret Manhattan Project, Feynman reduced long calculation timelines by inventing a technique called parallel computing. Two decades later, he won a Nobel Prize for creating a computing shortcut and birthing a new field of physics research called *quantum electrodynamics*. In the eighties, he was placed in charge of uncovering the cause of the *Challenger* space shuttle disaster and tried something his predecessors had not attempted: he spoke directly with the engineers. All of these accomplishments and more are dazzling enough, but we owe Feynman a greater debt: a hundred years from today, he will most likely be remembered as the father of quantum computing (*QC*).

Like its conceiver, quantum computing is complex and enigmatic. Feynman introduced the idea of quantum computing to the world in a keynote entitled "Simulating Physics with Computers," published in 1981. Four years later, he took on the more difficult task of designing one in a paper entitled "Quantum Mechanical Computers." Around the same time, computer scientist David Deutsch was attempting to use quantum mechanics as a stronger form of computer theory, called the *Church-Turing thesis*. Without going into the details, he wanted to know what it would take to define a computer capable of simulating and computing anything, including the physical universe. These are but two of the many people involved in the nearly four decades of work in making quantum computing feasible.

Quantum computing has taken on mystical quantities as the supercomputer of the future. This should come as no surprise.

Quantum mechanics is full of complex, mind-bending concepts like duality, superposition, and teleportation. Building computers to take advantage of these bizarre concepts feels a bit like unleashing fundamental physics to pick the locks of reality. It sounds cool, but *why* would we want this? What's the hype really about?

In the simplest terms, quantum computers can sidestep a barrier that plagues all modern computers (or as quantum computing experts politely call them, *classical computers*). As fast as today's computers may be at running calculations, they can only be in one state at a time. This is similar to how you are either sitting or standing at any given moment, but not both. This is called your state. The simplest state is one or zero, called a binary digit, or *bit* for short. Computers only know bits. A computer has a sequence of ones and zeros, perhaps trillions, but at any given moment, it is in a definite state, such as 1011100011001101...and so on. Classical computers calculate by changing states rapidly according to the instructions we call programs, and we pay people considerable amounts of money to write them to do things like record and share cat videos. The only way to make classical computers more powerful is to get them to change states faster or add more bits. But there are physical limits to both of these approaches.

Quantum computers, however, don't have the same limitations. They aren't bound to a single state at a time—they can be in all states simultaneously. We'll dig into the details shortly, but suffice it to say, when a classical computer has, say, 64 bits, it can have just *one of* 64 possible states. On the other hand, quantum bits, or *qubits*, can be all states at the same time. This means that a 64-qubit computer can be *all* 64 possible states at once. Mathematically, this is 2^{64}, or 18 quintillion possibilities. It could take hundreds of years for a classical computer to represent every state in a calculation, while a quantum computer can do this in a fraction of a second, a natural consequence of its structure. And this isn't entirely theoretical. In October of 2019, Google demonstrated that their custom 53-qubit quantum processor had executed a toy solution in 200 seconds—one that would have taken a classical supercomputer 10,000 years. This was the first claim of quantum supremacy, or the point where a quantum computer can run a program that is unfeasible for any classical supercomputer to match.

All of this is, to say the least, a simplification. We're still in the early days of quantum computing, and there are substantial limitations. But this may help you understand the stakes here. With a 300-qubit quantum computer, you could calculate as many possibilities as there are atoms in the universe. And we're closer to that day than you may have previously heard.

Until then, quantum computing remains an active area of research, development, and investment. There are handfuls of varieties and counting, and the type of quantum computer you use will rely heavily on the problem you're trying to solve. Are you modeling molecules for drug research? Try a quantum simulator. Do you need to optimize a supply chain or predict the weather? Then go with a quantum annealer. Are you looking to crack computer encryption schemes with Shor's algorithm? Then go with a universal quantum computer. Once you narrow down the class of problems you have, you have to decide between a handful of technical tradeoffs. Do you require long coherence time? Short gate operation time? More qubits? Is fault tolerance a requirement? Answers to questions like these will lead you toward one of many implementation options.

Of all the technologies that we've investigated in this book, quantum computing is the farthest out, likely to emerge around 2030. Jumping into quantum computing now is like getting involved in the internet in the late eighties, or the personal computer revolution in the seventies. Welcome to the quantum gold rush.

A QUICK APOLOGY

Before we dive deeper into this, I'd like to start with a *mea culpa*. If this topic we're about to embark upon is new to you, it's rare to grasp it all in one sitting. That is no one's fault, and certainly not yours, dear reader. It's the weakness of our human cognition and therefore human grammar that the complexities of nature seem so unnatural to us. Evolutionary biologist Richard Dawkins calls the scale in which we humans (and thus our brains) evolved "middle world," where we are sheltered from the strangeness of massive cosmic scales full of oddities like black holes. We're also far from the very small quantum world full of quanta doing queer things like entanglement. While we do the best we can, fitting a quantum information, quantum mechanics, and quantum computing education into a single chapter is like trying to thread a very small needle. What I hope you may gain here is an appreciation of what quantum computers might do, the difficulties surrounding why they're not doing them yet, and why the world will change dramatically when they succeed.

BENEFITS

Quantum computers are powerful because they represent *doubly exponential* growth compared to classical computer's mere *exponential* growth. What this means is that each extra qubit doubles the memory space of the previous amount. Five qubits is double the power of four, and 302 qubits double the power of 301. For a classical computer, the only way to double its power is to double everything—components like memory, transistors, and clock speed. Which, thus far, they've been doing amazingly well. Classical computers have doubled processing power every two years for the past five decades, following something known as Moore's law. It's an exponential growth. Every two years, computers become twice as powerful as before, which is why you can buy a smartphone today with more processing power than a supercomputer from the nineties.

Quantum computers have their own growth pattern that's doubly exponential, called *Neven's law*. Doubly exponential means that rather than "doubling" every generation, it grows by an exponential power. For example, in seven generations, a classical computer can grow 128 times more powerful (calculated as "two to the power of seven"). Meanwhile, a quantum computer could grow 340 billion billion times more powerful in the same time frame (calculated as "two to the power of two to the power of seven," or in other words, "two to the power of 128"). We'll get into how this happens in a moment, but suffice it to say, they do it by hacking the very fabric of the universe.

The holy grail of computing is causing billions of dollars to flow between major corporations, academic institutions, and government agencies. This technology is worth trillions. In the near term, there's a race toward *quantum supremacy*, another name for the point where a quantum computer can win against the world's top classical supercomputers. Some companies prefer other metrics, like qubit count or quantum volume, but all of these are imperfect measurements. The true test will be the applications of quantum computers for practical purposes. The expected benefits are so great that the first to market with a general purpose universal quantum computer will steer medicine, physics, industrial, and even artificial intelligence for decades. That's why companies like Microsoft, Google, and IBM are involved, as well as a laundry list of smaller companies like D-Wave, IonQ, and Rigetti. There are also players you might not expect, like Amazon and Honeywell. World governments have also jumped into the fray, pumping billions into quantum computing research. In 2018, the US passed the National Quantum Initiative Act to build and fund a national quantum computing team.

The reason for all of this excitement goes beyond the academic. Some of the top candidate use cases ripe for disruption by this very deep technology are molecular simulation, optimization problems, security, and AI.

MOLECULAR

To channel an example by Immanuel Bloch, simulating the states of a mere 280 particle compound requires a computer capable of representing 2^{280} simultaneous states. The problem with building a computer this size is that there are fewer atoms in the observable universe. So we need another approach. The first use case for quantum computers suggested by Richard Feynman was to simulate molecules. The benefits of true molecular simulation are beyond measure, including chemical reactions, the discovery of new medicines, and entirely new materials.

Quantum simulators effectively work by letting nature take its course in a controlled environment. It's not much different than using the physics of a canon shot along the ground to model the movement of planets encircling each other like Isaac Newton did. It was a small-scale experiment, and its ground state represented something else that had to follow the same physical laws. While the idea is decades old, Google was able to simulate a hydrogen molecule in 2016, while IBM was able to simulate two molecules in 2017. A quantum computing company called IonQ calculated the ground-state energy of a water molecule in 2019. While these may seem like slow improvements, we must remember Neven's law. We'll be in the range of more complex molecules soon, which we'll need for modeling new medicines.

Pharmaceuticals tend to be molecules of under one hundred atoms, putting us within spitting distance of modeling medicines. By the end of the decade, we should have the holy grail of medical computing known as protein folding, which will allow researchers to simulate basically any kind of drug and its interactions with organic material. Currently, this kind of modeling requires vast resources. For example, in 2020, an effort went underway for people to donate exabytes and months of computing power to investigate drugs for one disease: SARS-CoV-2, aka COVID-19. In the future, these simulations could be run in a few hours in one lab.

As quantum computers grow larger and are better understood, we'll reach a point where molecules of arbitrary size and complexity can be simulated. This opens the door for new material investigations, from simple modeling, like graphene and borophene, to potentially complex investigations of high-temperature superconductors, safer fuel cells, and vastly superior batteries. If it's made of material, it can be improved. This benefit extends not only to the materials themselves but also to attributes like biodegradability and lower manufacturing costs. Simulating materials allows quantum computing researchers to model creation of new ways to build better next-generation quantum computers, bootstrapping and accelerating quantum computing progress.

OPTIMIZATION (LOGISTICS, FINANCIAL, WEATHER)

In early 2020, Delta Air Lines announced a partnership with IBM to optimize several aspects of the airline including, interestingly, logistics. Airlines have notoriously complex requirements, such as optimization of traffic patterns, fuel consumption, real-time rescheduling, and pricing. The value of this endeavor is expected to reach into the billions, not only by constraining costs but also by increasing revenue through improved consumer experiences. The class of mathematics necessary to calculate the best answers to these issues is called *optimization problems*—a class that quantum computers are well suited to solve. Optimization problems are so ubiquitous that quants and data scientists tackle them as a matter of course. When advertisers ensure the right consumers see the right ad, it's mathematically similar to a financial services company optimizing the management of an index fund.

Improving optimization has so many benefits that we could easily fill pages if we listed them all. There are benefits for distribution (capacity planning, fleet routing), production (machine allocation), investing (portfolio optimization), advertising (media planning), human resources (workforce scheduling), and even back-office workflow improvements (business process management). There are basic technology improvements as well, such as signal routing to minimize crosstalk, or frontline improvements, like ensuring the best customers are prioritized when they pick up the phone. And those are just a taste of the general improvements. Add in industry specific optimizations— like smart electric grid routing or factory farms optimizing crop yields to cost—and the benefits of real-time, large-scale optimization truly feel endless.

Considering their flexibility and value, constrained optimization solvers are among the most popular data science tools used by corporations today. But there's a catch. It turns out that true mathematical optimizations are difficult for modern computers to solve, so most of these classical software packages employ a series of tricks to find reliable results. They're difficult because they require astronomical numbers of variables and relationships, and even relatively small optimization problems can quickly overwhelm the capabilities of a classical computer. But quantum computers excel at holding many possible states. If we had to pick a single killer app for quantum computers to drive the biggest trillion-dollar returns, it would be QC, as a general purpose optimization solver.

Several quantum computing companies focus their entire business plans on these problems. In the same way molecular simulation is best solved with specialized quantum computers, called *quantum simulators*, optimization also has its own specialized machines. One such type is the quantum annealer (QA) championed by a company called D-Wave. Let me elaborate on how this machine works.

When shaping molten glass into the shape of a vase, you have to let the glass cool slowly. Cooling too quickly fractures the glass, causing it to break. So you place the vase into an annealer, which is like a reverse oven. It starts hot and slowly cools to room temperature throughout the day, giving the glass particles time to settle into a stable position. This is akin to how quantum annealers work. You represent a math problem called a *cost function* through interconnected qubits, and as you let the system "cool" to lower energy states, you will eventually reach the ground state. At this point, like magic, the cost has been minimized, and an optimal solution is presented.

QAs are a more specialized form of adiabatic quantum computer (AQC), which in short is a theoretical, universal QC that has yet to be built in practice. An AQC is exciting because such a device could efficiently solve optimization problems as well as any other quantum computation. On the flipside, it's also possible to solve optimization problems on a universal quantum computer algorithmically by way of Variational Quantum Eigensolver (VQE). Time will tell which approach will prevail, if either. But no matter the method, the value will be the same: optimizing an inefficient world.

SECURITY

It's easy to make treasure inaccessible to anyone—just throw it into a volcano. But the art of security is to lock your treasure away from everyone, accessible only to you when you need it. Authentication is the core of security. It turns out that if you base your security on the immutable laws of the universe, certainty is possible. One interesting rule of quantum physics is that entangled quantum particles can only be observed once. Think of them as the shyest things in the universe—they're like a teen when they're around their crush. If they get a mere look from them even once, it will change them forever. Say I have a bunch of quantum particles, and I encode a message with them. Then I send some paired particles to you, and you read them. Depending on the state, you can be certain when someone else peeked at the particles in transit. If they aren't in the form you expect, it means someone looked, and you simply ask me to try again. Eventually, I'll get a message to you, and only you, with relative certainty that no one else saw the message. How and where can this be useful? Let's imagine you're a bank, and I want to make a withdrawal. With this system in place, we can both be certain that a message is coming from me and not from anyone else.

This is the essence of encryption, where we both have keys to encrypt messages: we're sure that when I lock my message, it can only be unlocked by you. The problem with the classical computers of today is that if someone else can get or guess the key, there's no way of knowing whether or not the message has been read by an eavesdropper. But a quantum system will always betray an eavesdropper's presence.

Quantum Key Distribution (*QKD*) *is* a type of cryptography that uses quantum communication to exchange secure keys between parties intended to be used only once. This is the best of both worlds, quantum and classical. QKD can ensure the keys you and I share to encrypt our messages are known only to us, through the use of a quantum communication channel. While the QKD shares the keys, we'll still communicate through the internet we know and love today. This is powerful because we'll gain a type of quantum security without throwing away our existing infrastructure. It's unlikely anyone will guess or steal a key so rapidly, and by the time someone has potentially hacked our keys, our QKD channel will have already generated and traded brand-new keys. It's like the key to your house randomly changing form every five minutes. It wouldn't matter if someone copied your house key because you'd have a new one relatively quickly. Quantum Xchange sells a "quantum-safe key distribution" that can be used today. Larger network security companies like Cisco are also in on the action.

Beyond quantum communication channels and QKD, there's

another security benefit of quantum computers. Unlike classical computers, quantum computers can generate truly random numbers. Randomization in classical computers is, in truth, only pseudorandom. The large generated numbers necessary to make a secure key are somewhat predictable with the right tools, with techniques and tools as varied as guessing computer clocks or reading the power surges of a plugged-in desktop. Quantum computers, on the other hand, are truly random.

With all of the benefits of quantum security, the next step is building a series of quantum networks toward a fully quantum internet. We're in the early days of this, but there are already a handful of functioning quantum networks, including the DARPA Quantum Network in the US, the Tokyo QKD network in Japan, and the Beijing-Shanghai Trunk Line in China. There are already quantum satellites as well, and we're a few years away from these networks beginning to span countries and the world. This won't fix all of our communication security problems, but it's a big step toward solving the age-old problem of being certain that we are who we say we are.

AI

Long before there was artificial intelligence, there was the artificial neuron, or the *perceptron*. The perceptron became the building block of neural networks and, later, the deep neural networks that power the current AI revolution. Published in *Nature* in 2019 as "An Artificial Neuron Implemented on an Actual Quantum Processor," a team of researchers were able to create an artificial neuron on IBM's quantum computer. It's still the early days, but there is a lot of evidence that quantum AI is not only possible but inevitable.

What we didn't cover in the AI chapter is exactly how deep neural networks are built. They're designed as a series of tensors, or multidimensional matrices, weighted to fit successive inputs. This means they need to support a high number of variables and connections. Does this sound familiar? It's superficially similar to the needs of optimization problems, and quantum computers may be well suited for the class of problems that deep neural networks solve. One of the biggest problems of modern AI, just like optimization problems, is that the smarter the AI gets, the larger the model needs to be, and an exponentially larger classical computer is necessary to train and run the model. Is it possible to leverage quantum computing to short-circuit the growing computation requirements of AI? Some seem to think so. When Google and NASA created the Quantum AI Lab (QuAIL) in 2013, they used D-Wave quantum annealers in their first few attempts. They eventually moved to a series of custom-built machines with UC Santa Barbara before declaring quantum supremacy in 2019.

The jury is still out regarding quantum AI, but many of the world's leading AI researchers are convinced of its potential. Quantum AI would marry quantum mechanics—the most elegant and subtle expression of reality—with artificial intelligence, the greatest human tool to be designed. Beyond all the practical benefits of AI, birthing an intellectual power that is rooted in the deepest energies of the universe, rivaling and surpassing our own intellect, will have lasting impacts on future generations beyond what we can imagine.

A BRIEF INTRODUCTION TO QM

"I asked: 'What does mathematics mean to you?' And some people answered: 'The manipulation of numbers, the manipulation of

structures.' And if I had asked what music means to you, would you have answered: 'The manipulation of notes'?"

—SERGE LANG

Learning quantum computing is more than math and code; it's also couched in the language of particle physics. If it were possible, I'd happily jump right to the meat of how quantum computers work, but first, we need a map, which is this brief overview of quantum mechanics.

We'll try and cover this topic with as little background physics as possible. Drilling into concepts like quantum electrodynamics, string theory, holography and so on, while interesting, constitutes entire fields of research, endless books, and deeper mathematical distinctions than necessary for our purposes. That said, while we should keep this simple, some ideas are so fundamental to quantum computing that they cannot be skipped, namely the following: quanta, duality, interference, superposition, spin, and entanglement. Don't worry about the jargon. Many of these concepts are weird but not too difficult to understand when tackled one at a time.

QUANTA

To understand quantum mechanics, it helps to understand atoms, the fundamental building blocks of all matter. To start, negative electrons orbit around a positive nucleus. If you think of the solar system, the sun is like the nucleus, and the planets are like electrons. When an atom gains energy, such as by heat, the electrons can move to a higher orbit. When they lose energy, they drop to a lower orbit. The first bizarre feature of quantum mechanics is that these electrons do not move smoothly between higher and lower orbits. When an electron changes orbits, it jumps to a higher or lower position. There is no in-between. This jump is called a *quantum leap*. The discrete energies involved are called *quanta*. Imagine Mars hopping into Earth's orbit without traveling the space in between, just disappearing from the Mars orbit and reappearing alongside Earth. It sounds like science fiction or fake news, but fundamental particles are always doing this.

The nature of discrete quanta versus the continuous energy that we're used to seeing in everyday objects like baseballs is what gives quantum mechanics its name. There are technical physical reasons and endless experiments and theories behind why we know this to be true. But for our purposes, it's enough to know that fundamental particles hop around in a quantum way. It's weird but it's true.

DUALITY

Fun historical fact: Einstein did not win a Nobel Prize for his most famous equation $E = mc^2$, nor for his theory of general relativity. This particular honor was awarded to him for showing that light is both waves and packets of energy called *photons*, work that was later generalized by Nobel Prize winner Louis-Victor de Broglie to include all matter. The packets of energy-like light and electrons are called *fundamental particles*—or, as we already know, quanta.

It may be tempting to assume that particles are simply weird dots of energy that always jump around in space. If only it were that simple. Yes, they are packets of energy. But they are also waves, like water waves. They are essentially both in a phenomenon called wave-particle *duality*. Like quantum leaps, there are plenty of experiments showing that this is true, even though it bends common sense.

One of the earliest popular experiments showcasing the weirdness of the quantum world is known as the double-slit experiment. The surprising results are akin to two blindfolded people taking turns throwing darts at a dartboard, one at a time. But rather than seeing a scattered plot, we instead find a distinct pattern of lines, like two interfering waves. But they're just throwing blindly, so how could this possibly happen?

The double slit experiment shoots quanta, such as photons, through two slits in a panel and then looks at the wavy pattern it creates behind it. The pattern is caused by wave *interference*, like the ripples caused by dropping two pebbles onto the surface of a still pond, rippling outward as a series of concentric circles. The ripples naturally interfere, making a lovely pattern on the surface. Since fundamental particles are waves, a similar pattern emerges on the other side of the slits. The plot twist is that this interference pattern appears even if we shoot particles one at a time. Think about that for a second. With only one particle, what exactly does it interfere with? The answer is, itself. The shot particles have properties of discrete particles of energy as well as waves. They are both. This is known as wave-particle duality.

It may seem like I'm harping on this point, but it's core to understanding quantum computers, so it bears repeating. If you remember only one thing, remember this: *fundamental particles, or quanta, act as waves until they collapse at a specific point in time.* What causes this collapse to happen? You may be shocked to find that it occurs once we observe or measure it. When unmeasured, subatomic particles act like waves on the surface of water. But only when we look does the pattern emerge. Observing the wave causes it to collapse into a specific point, and then the wave function acts like a particle.

SUPERPOSITION

So far, we have learned that the universe is made up of fundamental particles called quanta that exhibit wave-particle duality. And when we observe the wave function, it collapses into a definite particle at a definite place and time, like a point on a dartboard. We know that particles are also waves because they create interference patterns.

You may wonder where the particle is before the wave function collapses. Where is the photon when the light is still "acting" like a wave? That is an important question, and the answer is: potentially at all places on the waveform. The possibility of a particle being at multiple places at once is called *superposition*.

It's hard to gain an intuition of superposition because it's not something we encounter every day, but here's an analogy that may help. Have you ever felt *poignancy*? A moment that can't be described as a positive or negative emotion, just...important? We could call poignancy a superposition of emotion, not happy or sad *per se*, but both. A mixed emotion. Only upon further reflection can we perhaps more clearly describe how we felt about that event, positive or negative. The emotional wave function of, say, high school graduation has collapsed, and you were happy overall, though you may have cried at the time.

In the regular positions we are used to in our macro world, things are at a certain place at a certain time—like right now, a coffee mug is in the upper left portion of my desk. If the mug were in a superposition, it could be everywhere on my desk at once. But once I search for the mug, it collapses to a definite location. Fundamental particles exist in a state of superposition until they're observed, and then they collapse into a single place in time. And the important thing to note is that the exact location of the collapse is always random, within a probability zone. Where the particle eventually collapses is not merely unknown, it's *unknowable*, following a proof called Heisenberg's uncertainty principle.

This idea so disturbed early physicists that one created a famous thought experiment: Schrödinger's cat. Erwin Schrödinger postulated that if superposition were true, you could place a cat in a box with an atomic device that randomly releases a particle. If the particle were in one state, it would kick off a reaction to release cyanide and kill the cat. If the particle were in another state, the reaction would not take place, and the cat would live. Keeping the box closed and unobserved would mean that, due to superposition, the cat would be both alive and dead until someone opened the box to look. In other words, the cat would exist in a superposition of life and death—not as a zombie cat, but there would be two simultaneous states of being for the cat. Once the state of the particle was observed, the universe would then collapse into a state where the cat would definitely be alive or dead. Luckily for cats everywhere, no one has actually performed this experiment, and luckily for quantum physicists, this is an oversimplification. But it should give you some comfort to know that superposition is so strange that it continues to confound even the brightest minds in the field of theoretical physics.

SPIN AND UNCERTAINTY

"God does not play dice with the Universe."

—ALBERT EINSTEIN

"Don't tell God what to do with his dice."

—NIELS BOHR

Spin is angular momentum of a quantum particle, like an electron. Think of a top spinning clockwise on the floor (aka *up* spin) or counterclockwise (aka *down* spin). In 1928, a physicist named Paul Dirac showed that spin was a natural consequence of quantum physics through his eponymous Dirac equation. (And remember the name Dirac, as we'll revisit him soon).

Like quanta location and velocity, spin also exists in superposition until measured, and then it collapses into a definitive up or down spin. That's like our top spinning both clockwise and counterclockwise at the same time in a dark room, but once we turn the lights on, it picks a direction. The top no longer exists in a superposition; it collapses into a definite spin direction. This bizarre property is a fundamental component of quantum computing.

I like to think of spin superposition akin to the classic conundrum: If a tree falls in a wood and no one is around to hear it, does it make a sound? In quantum mechanics, this would be: If a particle spins and no one is around to observe it, how does it spin? The answer, of course, is uncertain.

ENTANGLEMENT

"Spooky action at a distance" is what Einstein called quantum entanglement when he first learned about this phenomenon, which is as good a description as any. So far, we've focused on single quantum wave-particles spinning in superposition, but the universe gets weirder when particles interact with each other.

Entanglement is one of the most important concepts to grok for quantum computations, and everything we've discussed up to now leads to this. Entanglement, also known as the Einstein-

Podolsky-Rosen (EPR) paradox, is a peculiar situation in which two particles interact and become connected. When one particle is observed, you can be reasonably certain about the state of the unobserved particle. If the spin of one particle is observed to be up, you would know that the second entangled particle spins down without even reading the second particle. More weirdness comes into play because this connection happens no matter the distance. Once two particles are entangled, reading the spin of one will instantly affect the spin of the other even if the particles are physically separated by a great distance faster than light. Hence the paradox, because nothing can travel faster than light. Nothing, it seems, except for the information exchange of entangled particles.

This example might help you better understand this. Imagine we have two spinning tops in a dark room, both spinning in states of superposition up and down, meaning clockwise and counterclockwise simultaneously. We place the two tops together in an entangled state and move one of the tops into another dark room. We observe one of the tops, causing the superposition to collapse into an up spin. Due to this, we can be reasonably certain that the other top will collapse into a down spin state, even without observing it.

Like many things quantum, though entanglement defies logic, it has been tested. A paper published in 2016 entitled "Quantum Experiments at Space Scale" by Chinese researchers showed entanglement working over 10,000 times faster than the speed of light by entangling particles and shooting a twin into space, and then measuring both of them. Other than practical limitations, there's no reason we couldn't test this at any distance in the universe. The ability to know the state of an unobserved particle by measuring the state of an entangled twin is an important component for implementing a universal quantum computer.

QUANTUM WRAP UP

That was a lot to absorb, and if this is your first experience with quantum mechanics, it can be a heady topic. Let's revisit these concepts one more time, tying them together into the necessary building blocks of quantum computation.

At the smallest scales, everything is made up of discrete packets of energy called *quanta*, which exhibit wave-particle *duality*. We're always uncertain as to exactly where a specific particle will be at any moment because it exists in a field of probability called *superposition* before the wave function collapses. A particle has *spin* as well, but like location, it spins in both directions—*up* and *down*—until we measure or observe it, and then it collapses into a definite spin. Particles can also *interfere*, meaning the states of some particles affect each other. Another kind of interaction is called *entanglement*, where the state of a pair of particles can always collapse into correlated states, like opposite spins.

With quantum basics as a starting point, we're now ready to jump into exactly what makes quantum computers perform magic.

TECHNOLOGY

As a quick reminder, quantum computers are interesting and valuable because they can execute at scales far beyond that of the greatest classical supercomputers that could ever be built. Richard Feynman's initial use case for quantum computers was to model quantum physics and chemical reactions in a way that could never be solved in a classical way, also known as quantum supremacy. This large-scale computation is possible because quantum computers can hold and act on 2N simultaneous states, compared to classical 2N states of only one at a time. This gives quantum computers a massive speedup.

But how does a quantum computer hold simultaneous states all at once? And how can we get a meaningful answer out of the QC? We'll focus on this in the next few sections.

THE DIVINCENZO CRITERIA

Before we jump all the way into the deep end, let's make a quick pit stop to clarify a common question: yes, there are many ways to build a quantum computer. However, we are still in the early days, so many research dollars are being spent on nascent material investigations. Some of the technologies being considered are physical qubit modalities like electrons or nuclear spins, photons with polarization, trapped ions, superconducting qubits, and more.

We determine good candidate technologies for quantum computer implementation based on a list of seven criteria designed by David DiVincenzo in the year 2000, colloquially known as the DiVincenzo criteria. The first five criteria are needed for quantum computation:

- A scalable physical system with well-characterized qubits
- The ability to initialize the state of the qubits to a simple fiducial state
- Long relevant decoherence times
- A "universal" set of quantum gates
- A qubit-specific measurement capability

The last two are needed for quantum communication:

- The ability to interconvert stationary and flying qubits
- The ability to faithfully transmit flying qubits between specified locations

What each of these mean can fill chapters among themselves. In the shortest terms: to build a quantum computer, we need stable qubits we can encode logic into and then read the output. To communicate quantum information means we need to encode and transmit qubits.

As long as the technology supports the DiVincenzo criteria, we can call a computer quantum, even if those effects are realized by macroscale, superconducting materials. As far as qubits are concerned, quantum is as quantum does.

QUBITS AND SPIN

A qubit is a quantum bit. Unlike a classical bit that is either a one or a zero, a qubit is, in the simplest terms, both a one and a zero, or a superposition of any state in between. This can be represented by an electron in a down or up spin, or a superposition between down and up spin.

A qubit representation is written in *Dirac notation*, also called *bra-ket*, pronounced "bra-x" ($<x|$) or "ket-x" ($|x>$). In short, it's an index vector notation. If you're unfamiliar with vectors, you can think of them as arrows aimed at a specific point. It's a convenient way to calculate a field of probable outcomes. Some example qubits in Dirac notation are:

$|\psi\rangle$: *A wave function*

$|0\rangle$: *Ket-zero, or up spin*

$|1\rangle$: *Ket-one, or down spin*

$$\frac{|0\rangle + |1\rangle}{\sqrt{2}}$$: *A qubit in superposition*

The math here isn't important for our purposes. Just note that you'll often see qubits represented in Dirac notation as a wave function in the Greek letter psi, or ket-psi $|\psi>$.

THE BLOCH SPHERE

Making sense of the Bloch sphere requires a general understanding of quantum mechanics, linear algebra, trigonometry in polar coordinates, and the geometric view of complex number fields. It also doesn't hurt to have an understanding of innate concepts in other disciplines as broad as probability, limits, and for good measure, intuitions, and intuitions like Euler's number. Speaking as a nonmathematician who's never taken a single physics class, it's mind melting to say the least. That's the bad news.

The good news is that with a handful of concepts, programming a universal quantum computer can be done with the help of the Bloch sphere and some rules of thumb. The Bloch sphere is a 3D representation of states that a particle's wave function can take, thus making a useful model for qubits.

Recalling geometry, a sphere is a three-dimensional object like a beach ball or a globe. All 3D objects can be represented with three-axis coordinates, and a Bloch sphere is no different. If you imagine a line drawn right through the north pole to the to south pole of the sphere, it's called the Z axis. There are also perpendicular X and Y axes.

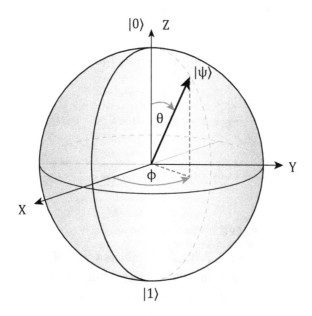

A Bloch Sphere with X, Y, Z coordinates.

In a basic case, if a particle spins up representing a qubit |0>, it's represented as a point at the north pole of the sphere, at the very top. Meanwhile, a |1> is the point at the south pole.

The north pole is 0, the south pole is 1, and everywhere else is a probability between them.

Any point on the sphere in between |0> and |1> is in a super-position state. The superposition value that we saw above (|0> + |1>) / sqrt(2) can be represented on the Bloch sphere at the equator.

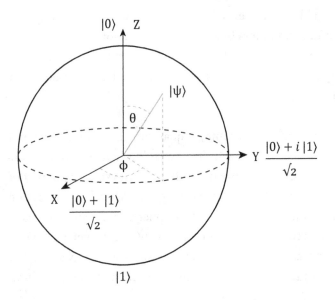

Superposition on a Bloch Sphere on the equator.

This value, along with its negative ((|0> - |1>) / sqrt(2)) is another axis called the X axis. For completeness, the third dimension is the Y axis representing something called phase, which we don't need to worry about here.

QUANTUM LOGIC GATES

What we've done so far with the Bloch sphere is show how a quantum computer can hold onto multiple states at once by put-ting a qubit into superposition. In a classical computer, it gets interesting once we start performing operations on our bits—it's operations that allow a collection of ones and zeros to perform math and, eventually, program AI to drive autonomous vehicles.

What we need are quantum logic gates to run calculations on qubits. Unlike classical operations, qubits must adhere to physical laws like reversibility. To match the needs of qubits, quantum computers have different operations from the basic AND, OR, and NOT gates of classical computers. Instead, you'll see quantum gates with various names like Pauli-X, Hadamard, and CNOT.

The details of programming a quantum computer are beyond what we need to know in this book, but here are some examples to give you a sense of familiarity. The simplest family of quantum operations are the Pauli gates. What they do is rotate values around the axes of a Bloch sphere. An X gate is the most common rotation, akin to a classical NOT gate which converts a *0* to a *1*, and *1* to *0*. In the quantum world, a Pauli-X converts a |0> to a |1> and vice versa. There are similar rotations around the Y and Z axes as well.

$$|0\rangle = \begin{bmatrix} 1 \\ 0 \end{bmatrix}$$

$$|1\rangle = \begin{bmatrix} 0 \\ 1 \end{bmatrix}$$

Ket-zero and ket-one are just simple vectors.

If you want to get geometric about it, these actions are also called a π-pulse (pi pulse) since it's a half circular rotation, aka π. (Remember in geometry class, measuring around the circumference of a circle is 2π times the radius? Well, half of that is just π.) Many QC actions are described in pulses, like π/2-pulse around the Y axis, or π/4-pulse around the Z axis. While the Bloch sphere is a useful visualization, mathematically the Pauli gates are unitary matrix operations on the qubit. A matrix is like a vector, but rather than just a column of numbers, it also has rows in a rectangular form.

$$X = \begin{bmatrix} 0 & 1 \\ 1 & 0 \end{bmatrix}$$

$$X|0\rangle = |1\rangle$$

$$\begin{bmatrix} 0 & 1 \\ 1 & 0 \end{bmatrix}\begin{bmatrix} 1 \\ 0 \end{bmatrix} = \begin{bmatrix} 0 \\ 1 \end{bmatrix}$$

Performing a Pauli-X (or CNOT) operation flips a |0> qubit to |1>.

The purest example of a unitary quantum operation with no classical analog is the Hadamard gate, represented as an H. An H puts a |0> or |1> qubit into a superposition. In the Bloch sphere, this moves the vector from a simple north or south pole along the Z axis to a point along the equator between both of them. Since it's sitting exactly between the |0> or |1> poles, it has equal odds of collapsing into a definite one or zero when it's measured, which is the very definition of superposition.

$$H = \frac{1}{\sqrt{2}} \begin{bmatrix} 1 & 1 \\ 1 & -1 \end{bmatrix}$$

$$H|0\rangle = \frac{|0\rangle + |1\rangle}{\sqrt{2}}$$

H applied to ket-zero is a superposition between ket-one and ket-zero.

The second family of quantum gates are two qubit gates. A two-qubit system can be represented as |00>, |01>, |10>, |11>. Two qubits have four computational basis states. If these two qubits are in superpositions, they can hold four potential states of information simultaneously. Technically, there are amplitudes (alpha, α) assigned to each ket, and the four states are terms of a single equation |ψ>.

$$|\psi\rangle = \alpha_0|00\rangle + \alpha_1|01\rangle + \alpha_2|10\rangle + \alpha_3|11\rangle$$

$$|\psi\rangle = \begin{bmatrix} \alpha_0 \\ \alpha_1 \\ \alpha_2 \\ \alpha_3 \end{bmatrix}$$

$$|\psi\rangle = \begin{bmatrix} \alpha_0 \\ \alpha_1 \\ \vdots \\ \alpha_{N-1} \end{bmatrix} = \sum_k \alpha_k|k\rangle$$

A bit more detail for completeness, the same as before, just with a new variable.

This is the crux of the quantum speedup, calculated as 2N states, where N is the number of qubits. If we have a fifty-qubit system, we can hold one quadrillion states at once. One hundred qubits? Try one thousand billion billion. A 266-qubit computer can represent one variable for every atom in the observable universe at once.

But back to two qubit gates. In classical computers, one of the most important gates is called "exclusive or," also known as XOR. It means "pick this or that, but not both." A similar quantum variant is called "controlled not" or CNOT (sometimes also written as cX). The first qubit is the control qubit. If it's zero, don't do anything. If it's one, flip the second qubit around the X axis. Like the various Pauli gates, there are controlled Z (cZ) and controlled Y (cY) gates as well. This control structure has two purposes. The first is that it allows for a form of branching logic, analogous to an "if else" in classical computers. Second, qubit operations like the controlled gates can also entangle the two qubits, meaning the result of one will correlate with the result of the other in some way. For example, if we wanted to entangle two qubits so that reading qubit A as |1> would imply qubit B was also |1> and vice versa, we could apply a Hadamard operation to both qubits, placing them in superpositions. Then, we'd apply CNOT with A as the control, and B as the NOT operator. That simple chain of operations creates one qubit in superposition, the random result of one or zero and two that are entangled.

Like unitary gates, we can represent controlled-gate operations as a matrix that transforms the input qubits into a different output. And because it's a quantum computer, every state in the matrix can be represented at once to an enormous size before collapsing down to a single answer. That's the real magic of a universal quantum computer.

CIRCUIT REPRESENTATIONS

Between the family of one and two qubit gates that we saw in the examples above, we can theoretically code any program to be executable by a universal quantum computer. We saw representations of logic gates as names and equations as Dirac notation or matrix multiplication. But constructing a program as an equation is tedious. So another way of building a program is via sequence diagrams, starting as parallel lines, with each representing a qubit value on the left and moving to the right over time. Think of it as sheet music, where each bar moves progressively to the right and contains notes to play.

Consider a few of the gates we've discussed so far: Pauli-X, Hadamard, and CNOT. The first two are represented, respectively, as an X or an H with a square around them. CNOT, operating on two qubits, spans two lines with a dot on the controlling qubit, and a vertical line connected to a plus inside a circle which represents the affected qubit. Another important symbol is called *measurement*, which collapses the qubit into a final value. With operations in hand, putting qubits together on a sequence creates a quantum program.

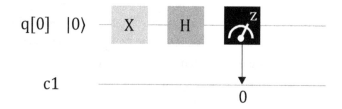

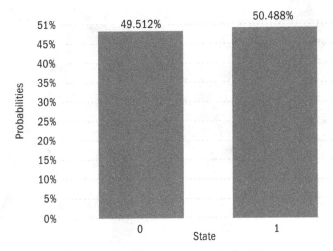

The likelihood of a superposition collapsing to one or zero is about 50 percent.

Some systems, like IBM Q, provide tools for building a sequence representation and running it on quantum hardware. Here's an example of a quantum program on IBM Q mentioned previously. It uses Hadamard and CNOT gates to place a qubit into a superposition and entangles the qubits so that measuring one will imply the other.

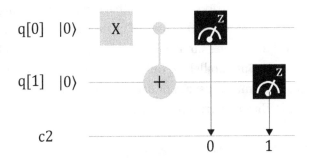

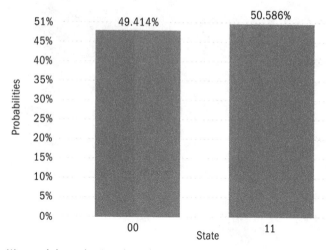

We entangled two qubits. Note the results are 00 or 11, but not 01 or 10.

This program takes advantage of two important core concepts to programming a quantum computer: parallelism and interference. One has been implied thus far; the other is new.

PARALLELISM AND INTERFERENCE

Because of quantum superposition, quantum computers can hold many possible states all at once. You've seen this represented as huge matrices of numbers that can easily grow bigger than the total atoms in the observable universe. The ability to hold all of these states simultaneously via superposition is called *parallelism*. But parallelism only gets us so far. How can we collapse this massive space of data down into an answer that we need? This is where *interference* comes in.

Two or more waves can either amplify or cancel each other out in a process called constructive or destructive interference. Think of hearing aids and noise-cancelling headphones. Hearing aids amplify sound waves to make them louder, or constructive. Noise-cancelling headphones do the opposite, emitting opposite waves to dampen the ambient waves and thus the sound, or in other words, they are destructive. By carefully controlling interference of quantum bits so that some values are cancelled out while others are amplified, the output of our quantum computation can land on a single correct solution when at the right magnitude.

The simplest interference is two consecutive Hadamard gates. The first H gate places the qubit into superposition, so when it's read, the output is a 50 percent random one or zero. The second H gate causes an interference of potential outcomes, causing the value to revert back to a zero output.

Not all quantum algorithms rely on interference, but some famous ones like Deutsch-Joza and Shor's algorithms do. We'll visit Shor's algorithm and its implications for global security soon.

HIGH-LEVEL LANGUAGES

Circuit representations are useful for small programs, but high-level programming languages are a necessary step toward industrial adoption of quantum computation. We've looked at quantum instruction sets, but there are many other useful quantum executions that are quite complex, such as Quantum Fourier Transform (QFT) that could be a single line in code.

Here's an example of code for Quantum Teleportation in Microsoft's Q#.[5] While learning any new programming language

---

5   "Quantum-Code," GitHub, accessed February 11, 2021, https://github.com/amarchenkova/amarchenkova-quantum-code/.

takes effort in its own right, this is somewhat more readable than a collection of symbols and lines.

```
/// # Input
/// ## msg
/// A qubit whose state we wish to send.
/// ## target
/// A qubit initially in the |0> state that we want to
send
/// the state of msg to.
operation Teleport (msg : Qubit, target : Qubit) : Unit
{

    using (register = Qubit()) {
        // Create entanglement we can use to send the
message

        H(Q_Alice);
        CNOT(Q_Alice, Q_Bob);

        // Encode the message into the entangled pair,
        // and measure the qubits to extract the
classical data
        // We need to decode the message into the target
qubit

        CNOT(msg, register);
        H(msg);
        let data1 = M(msg);
        let data2 = M(register);

        // decode the message
        if (data1 == One) { Z(target); }
```

```
        if (data2 == One) { X(target); }

        // Reset our "register" qubit before releasing
it
        Reset(register);
    }
}
```

IBM, Google, Microsoft, QCI, D-Wave, IonQ, and Rigetti are all working on quantum computers. Some of these companies have several kinds of hardware and different languages for various layers. IBM Q's Circuit Composer has QASM, Microsoft has Q#, and D-Wave's Ocean has SDK, and so on.

There are as many companies as there are languages building quantum computers. The company you choose depends on the hardware you are using, the level you're operating at, and the programming methodology you're comfortable with.

We're in the Wild West of quantum computing research. While we've cleared the Oregon Trail, we still lack much of the infrastructure necessary to move at top speed. But rest assured, the railroads, laws, and cities are on their way.

PROBLEMS

There's an old joke I like to tell about an Irish farmer. A passerby asked, "How do I get to Dublin?" The farmer took off his hat, shook his head, and replied, "Well, if I were you, I wouldn't start from here." That sums up the question of "How can I use a quantum computer?" Don't start at quantum. Start with the problem, and when all other options are exhausted, only *then* consider a quantum computer. There are many obstacles to conquer before quantum computers can catch up to classical computers. We have technical limitations, implementation complexities, as well as security issues. Finally, there's the tipping point of knowledge and tools required to break into popular usage.

TECHNICAL ISSUES

The current technical limitation of modern quantum computers can be rolled up into a single acronym *NISQ* (noisy, intermediate scale quantum computer). "Noisy" means that coherence times are low and error rates are still high, while "intermediate scale" means we need more qubits with more complex connections.

Coherence and Gate Fidelity

One of the problems in building a computer based on quantum mechanical effects like superposition is that qubits can only remain in that state for a brief window of time. That window is called *coherence*. All quantum computers must execute all operations within that coherence window before the wave function collapses, or *decoheres*. Some of the best systems to date decohere within a few seconds. Another complication is that the more operations take place, the more the chances increase for the qubits to decohere. In mid-2020, the Chicago Pritzker School of Molecular Engineering made great strides by extending coherence time by over 10,000 times. Improvements like these are cornerstone efforts to more practical quantum computers.

Coherence time is important, but even more so are the number of operations that can be executed properly before decoherence.

This can be measured as *gate fidelity*. There is plenty of work happening in the pipeline to improve gate fidelity, from longer coherence times to shorter gate operation time to increased gate accuracy. In mid-2020, Honeywell claimed the best quantum computer by a metric known as *quantum volume*, which effectively runs the quantum computer against a series of algorithms to come up with a number based on the number of qubits, error rates, coherence times, and more. While the quantum computer itself was low qubit, the number of qubits is only one of many competing metrics.

Some good news is that while we live in a world of NISQ, there's a field of research into Quantum Error Correction (QEC) that can accept and correct for various quantum fluctuations. The downside is QEC requires even more qubits.

More Qubits

"The Google plan is roughly to build a million-qubit system in about ten years, with sufficiently low errors to do error correction."

—JOHN MARTINIS

In addition to gate fidelity, the other large technical problem is the "intermediate scale" part of NISQ. Many of the most interesting longer-term use cases that require quantum computation, from modeling nontrivial molecules to cracking encryption, need far more qubits than we currently have on hand. And there is plenty of work being done to build quantum computers with more qubits.

The top early contenders this decade in the high-qubit space seem to be Google, who claims to be on the path to build a million-qubit quantum computer by 2030, and a startup called

PsiQuantum, which was funded to the tune of $215 million to build one million qubits "within a handful of years." While we can hope they're on the right track, only time will tell. On a shorter time horizon, IBM claims they'll have a 1,000-qubit QC by 2023.

Another alternative to more qubits is something called qudits (with a *d*). As we know, quantum particles are complex and hold many kinds of information. We talked about spin up and down, but there are other measures like energy state, photon polarization, and wavelengths. It turns out some particles can be entangled in a high number of dimensions far beyond the two of up or down. The National Institute of Scientific Research in Canada was able to entangle photons in a superposition of ten wavelengths, thus creating a ten-dimensional-qudit. Joseph Lukens at Oak Ridge National Laboratory told *IEEE Spectrum*: "I do think a 96-by-96-dimensional system is fairly reasonable and achievable in the near future." If this emerging research is correct, many of our estimates about the power of quantum computers may fall short. Two 96-dimensional qudits have more data density than thirteen two-dimensional-qubits. Suddenly, we won't need to deal with the complexity of building a 1,000-qubit computer to calculate complex proteins. Instead, we'll just need a mere 152 qudits.

IMPLEMENTATION ISSUES

When the light bulb was being invented, Edison's lab experimented with different filaments that glowed long enough before burning out. They weren't sure exactly how to build a light bulb, but through numerous experiments, a candidate emerged. Soon, competitors made their own attempts. Over the years, we were presented with choices for electric lights, from incandescent to excited fluorescent gas to light emitting diodes. There's some similarity with quantum computers.

New research is released weekly, and some will drastically change the direction of this nascent technology. It's still early in the race, so declaring a winner is premature, but there are horses worth betting on. Superconducting qubits in particular are popular thanks to some of the leading groups, namely Google, IBM, and D-Wave. These are not quantum per se, but circuits that fit the DiVincenzo criteria by controlling quantum effects through a Josephson junction. Artificial atoms, quantum dots, and linear ion traps are also qubit candidates. Majorana fermions is one of the newer options being championed by Microsoft.

Each technology has its strengths and weaknesses. Some quantum implementations tend to be highly sensitive to noise from the environment, while others are quite resistant. Some have long coherence times but slow or unreliable gate operations. Others decohere quite rapidly. In the case of quantum annealers, they still haven't been mathematically proven to operate in a quantum way. And in almost every case, most quantum computer implementations must be kept very, very cold.

Refrigeration

Every so often, you'll see an interesting headline, such as how Intel shrank quantum chips to fit in your hand. These headlines make it easy to imagine a near future of handheld quantum devices. The biggest blocker to such possibilities is that today's quantum computers still take up a fair bit of room. It's not the chips that need all that space. It's the refrigerator.

The quantum world, as you know by now, is very strange. It

operates at such complexity and speed that it's difficult for us to harness it the way we need. So our only solution is to remove as much complexity as possible and isolate a handful of quantum artifacts. This means taking out all extraneous particles (vacuum) and energy (heat), and cooling quantum chips to near absolute zero. The current solutions are known as dilution refrigerators. They often look like a steampunk instrument—hundreds of pounds of nested brass drums, pipes, and wires generating hypnotic clicking and hissing sounds. Each drum gets smaller than the one above it, like a chandelier, with each layer getting progressively colder, down to the lowest container that houses a small quantum chip. Inside, it's colder than even outer space, and it's where the magic of quantum computation happens. As a quantum physicist friend says, "Building a linear ion trap is something grad students can do. Refrigeration is hard engineering." The good news is that refrigeration experts are getting into the game. Honeywell, a well-known HVAC company, has quickly moved deeply into the quantum computing space, flexing their refrigeration muscles.

SECURITY

Imagine a world where modern cryptography was broken. Every email you've ever sent could be read, your bank account could be accessed, government and military systems are easily accessible by anyone, and corporate IP and other secrets are exposed. Encryption is the basis of trust in the internet, validating that only approved parties are part of a communication or transaction. In terms of security, quantum computers giveth, but they also taketh away.

We see security as a benefit of quantum computers, but they also blow a hole right in the middle of current cryptographic schemes based on public key signatures, also called asymmetric keys. MIT Professor Peter Shor's algorithm can theoretically break asymmetric cryptography based on prime factors, such as the popular RSA algorithm that protects much of the world's secrets. Some estimates put the odds of a quantum computer capable of breaking current high-end RSA crypto schemes at 50 percent by the year 2030. If you're a government, military, bank, corporation, or private citizen, that risk is too high.

Shor's algorithm requires a universal quantum computer with thousands of qubits. While you might be tempted to believe you're safe because this computer is still years away, anything encrypted today can be stored and accessed in the future. Perhaps tomorrow everyone will be using postquantum cryptography, and perhaps a quantum computer large enough to run Shor's algorithm won't be built for ten years, but everything encrypted yesterday (and earlier) will be up for grabs. There's a real need to keep current sensitive information secret into the future.

In August of 2015, the US National Security Agency announced that it "will initiate a transition to quantum-resistant algorithms in the not-too-distant future." By way of a National Institute of Standards and Technology (NIST) sponsored competition, a standard should exist by 2022. We need standards (e.g., via NIST) for government, healthcare, and the financial industry, and we need them as soon as possible. If you're a CISO or running security for any organization, it's imperative to begin adopting quantum-resistant security very soon.

EDUCATION TO A TIPPING POINT

Beyond the cost, technical complexity, and other open problems, one of the biggest barriers to the adoption of quantum computation is knowledge. The number of people in the world in 2020 who could reasonably call themselves quantum computation experts is in the thousands, compared to the tens of millions who program classical computers daily. Whether due to the complexity of quantum mechanics, the dearth of tools, or the lack of many algorithms and solid use cases, many of the blockers we have are conceptual and community based.

The good news is that there are attempts to change this. Google's X moonshot program is working on quantum software and algorithms. There's a focus on bringing quantum to the masses, including AI that can convert classical programs to quantum. MIT has many courses teaching quantum computation through their xPRO program, and Microsoft has paired with *Brilliant* to provide free online classes. This subject requires considerable effort to master, but luckily, the math can be understood with the right combination of YouTube videos and Kahn Academy.

Moreover, companies like IBM and D-Wave provide anyone access to their quantum computers via the web. I'm a big fan of coding QASM on IBM's Q for learning quantum algorithms, while D-Wave is excellent for supply-chain-focused, constrained optimization experiments. Google has Quantum TensorFlow, leveraging their successful machine learning toolkit to simplify the quantum computing space. For as complex as AI might be, for the moment, quantum computing is more so. Quantum/AI hybrids look to be in our future.

All in all, it's a great time to be alive! Quantum computation is within the grasp of anyone who cares to learn. I'll never forget the experience of the first time I coded and ran quantum teleportation. It was like hacking the universe with a quantum lock-picking kit.

COLLAPSING THIS CHAPTER

When the light bulb was being invented, researchers experimented with different materials for filaments. When the first ones lit up, researchers paid a premium, took on tremendous risk, had faith the light bulb would help build a better world—they wanted to be first. In doing so, they paved the way for mass adoption, and that's where we are in the quantum computing adoption curve. We're in the "punch card" phase of quantum computing, breaking down problems to the most detailed machine code because we lack the tools and concepts to operate at higher levels. But this doesn't mean they're useless. The first computers cracked the Enigma code and got humans to space. While the goal is a fault-tolerant universal quantum computer, we needn't wait for perfection for viability or commercial use.

If you've made it through this chapter, congratulations are in order. This is the most out-there technology we've covered in more ways than one. Yes, it's a few years from mass adoption and yes, it's bizarre to comprehend at this early stage. But moreover, the future of quantum computers is still being written, and for those who dream of founding the next Microsoft or Apple, now is the time for the next generation of titans to come of age.

Quantum computers are being investigated today with use cases encircling optimization problems, security, and artificial intelligence. Startups and researchers are being funded to the tune of billions of dollars by governments and corporations, and communities are cohering around increased education and sharing solutions. Inventing a new way of working is not easy, but the biggest impacts happen early, echoing throughout the community for decades. This is the emergence of a world of unfathomable efficiency, security, and super intelligence. There are opportunities everywhere. I, for one, am choosing to jump headfirst into the quantum world, the deepest of the deep tech. I hope to see you inside.

FURTHER READING

- *Quantum Computing for Everyone.* Chris Bernhardt
- *Dancing with Qubits: How Quantum Computing Works and How It Can Change the World.* Robert S. Sutor
- *Quantum Computation and Quantum Information.* Michael Nielsen and Isaac Chuang
- *Surely You're Joking, Mr. Feynman! (Adventures of a Curious Character).* Ralph Leighton and Richard Feynman

9

DENOUEMENT

"This is not the end, it is not even the beginning of the end, but it is perhaps the end of the beginning."

—WINSTON CHURCHILL

Political unrest, last-stage capitalism, pandemics, and other black swans may remain major forces in the 2020s. But when history about this time period is written, one or more of the technologies covered in this book will have played an outsized role in the story. Will blockchain democratize finance and bring about a more equitable century? Can AI and IoT help usher in a more sustainable circular economy? Will quantum computers predict and prevent future pandemics? How we wield these new tools will echo through generations—and we're the ones lucky enough to live during the time where we get to decide.

DEEP TECH, REDUX

This book was intended to be a brief introduction of the tech that will shape the coming decade. We focused on seven deep tech ecosystems. While there's some overlap between them— in fact, overlap is where the most interesting opportunities lie—we kept the technology chapters largely distinct. There is good reason to believe that each of these seven technologies are rare gems whose benefits will run far beyond incremental improvements in a field or two. As we discussed in Chapter 1, they are what economists call general purpose technologies (GPT), a label reserved for technologies that affect the global economy through the creation and adoption of a new way of working—for the technologies that impact the very core of our society, often seen historically through clear inflection points, pre and post such-and-such invention, be it a product, process, or new organizing principle.

To recapitulate, the emerging technologies of focus were:

- *Artificial intelligence*: One of the biggest opportunities in the deep tech space. Over the past decade, advances in hardware and big data have converged—along with *deep neural network* architectures—to give rise to many new AI capabilities trained with *machine learning*.
- *Extended reality*: The catchall term for digital realities that range from head up displays and *augmented reality*, to *mixed reality* to full *virtual reality*. The biggest similarities are the use of headsets to cover the field of vision and augment or replace what we see with digital replicas.
- *Blockchain*: Though initially built to manage *cryptocurrencies* like Bitcoin, blockchain has since evolved to cover use cases where two or more parties need to transfer value in a trusted way without the need for a third party. This generic blockchain concept is also known as *distributed ledger technology*.
- *Internet of Things*: One of the largest opportunities in deep tech. The core purpose of IoT is to connect the physical and digital worlds by way of sensors (understanding the world)

and actuators (making changes in the world). Big use cases are industrial IoT and wearables.

- *Autonomous vehicles*: A specialized type of autonomous robot targeted toward the transport of people and things from point A to point B (examples include self-driving cars). Removing humans from the transportation loop reduces cost and risk. It also has the potential to change the structure of cities.
- *3D printing*: A type of additive manufacturing where materials are deposited via computer numerical control. This opens up new kinds of designs, reduces complexity and materials, and allows the printing of shapes that are difficult or impossible by standard techniques.
- *Quantum computing*: The next generation of computation in general. Unlike classical computers that have doubled in power every couple of years, quantum computing promises to exponentially double in the same time frame. In short order, QC will tackle secrets of the universe that are impossible for today's computers to solve.

Taken together, these technologies are at the forefront of a new industrial revolution called Industry 4.0. Through this and changing consumer habits, these technologies will add a minimum of $50 trillion in new value to the gross world product over the coming decade.

As I mentioned in Chapter 1, we chose to focus on the how, why, and what of the technology itself rather than existing use cases so as to remove creative barriers. Hopefully by now, the why is clear, but if not, here it is: focusing only on the application of IoT and AI to logistics limits the thinking of someone in the logistics industry who may see new opportunities in other technologies, like XR or blockchain. True, people are using quantum computing for optimization, but how about for new kinds of creative video games or education? Aren't those use cases best discovered by those with a passion for video games or education? You don't need me or academics or a catalog of existing companies to creatively apply the technology once you understand it.

MORE AND MORE

"Books are never finished—they are merely abandoned." This endlessly misattributed quote applies rather well here. I would have liked to add more chapters, doubling the length of this book—but this would have also ensured it would never be published. So I stopped at seven chapters, but drawing the line here doesn't mean our education must cease. Had this book continued unabated, these are the technologies I would have included. They're all worth discovering and are all likely to have some impact this decade.

- Autonomous robots
- Drones and flying cars
- Commercial space travel
- Next generation materials (graphene, borophene)
- Nanotechnology
- Gene resequencing, CRISPR/Cas9
- Synthetic biology
- Xenobots (living robots)
- Sea-based tech (floating farms and cities, power generation)

The exciting reality of deep tech is that it's never finished. It's an infinite game. Like religion or fashion, there's always more to debate, more to do. Any one of them may become a general purpose technology in the remaining century. Hopefully this book has inspired you to learn more, not only about the big seven we've covered, but to keep an eye out in the deep recesses of R&D labs and startups for the next line of deep tech—and be ready to pounce.

CREATIVITY TO ACTION

"Innovation is seeing what everybody has seen and thinking what nobody has thought."

—DR. ALBERT SZENT-GYÖRGŸI, NOBEL PRIZE LAUREATE

Now that we've covered a spectrum of deep technologies, the real fun can begin. The next steps of your journey lie beyond this book: it will start with creativity and then will be about getting it done. Like Thomas Edison said, "Genius is one percent inspiration, ninety nine percent perspiration." This book, along with a deeper understanding in your forebrain and the brilliant, creative ideas that come to you late at night are the one percent. That's the fun part. Next comes the work. It takes effort to create something new and just as much to get people excited and comfortable with the idea. Whether your idea comes to life by way of personal effort, inspiring a team to work, or hiring experts to work on your behalf, a big portion of that effort goes into making it real. Here are some tips that have worked well for me over the years.

If you're going the startup route, you'll need customers long before you seek investors. Speaking as someone who has helped create and grow six startups, I'd recommend taking a lean approach, focusing on the outcomes that real people want rather than the technology. If you can leverage deep tech to deliver an outcome in a novel way that's considerably easier or cheaper for your end users, then you have an endeavor worth pursuing with little competition. There's an old adage that says that the world wants a better mousetrap, but that's fundamentally untrue. People don't want a better mousetrap; they want an environment that's free from pests. If you can deliver the outcome they want with an emerging, deep technology, the world will truly beat a path to your door. You can start by understanding your customer's needs—design thinking workshops are a good start. Prototype the experience. Build a minimum viable product (*MVP*), and incorporate real customer feedback as quickly as possible. Be agile and iterate to continually challenge your assumptions. Make adjustments based on real experiences. Keep your company slim, following the lessons of lean startups. As you grow your market, laser focus on your core customers first, the early adopters that just "get it." Following the marketing lessons of the bestseller *Crossing the Chasm*, find beachheads to work your way into mass market penetration. Never stop iterating, improving, and searching. It's not easy, but that's why it's the ninety nine percent.

That's all well and good for startups, but what if you're innovating within the confines of an existing organization? Like startups, you need to be customer focused. Those customers can be external consumers or internal business partners. No matter the audience, you want to deliver an outcome for them—or for yourself. Many people within large organizations are comfortable with their current way of working and will insist that only a few tweaks be made to their existing process. Or maybe they'll agree to some new software. As someone who has a handle on the cutting edge of deep tech, you come armed with possibilities they haven't imagined.

Successful organizations obsess about metrics, which are designed to measure and optimize existing processes and improve margins, or grow revenue. This was a good focus when running a factory in the 1940s. But in the modern age, existential threats, from shifting consumer demands to a relentless army of upstart competitors (who always seem to leverage technology better than you can) means the ability to survive and thrive requires increased flexibility. Processes such as agile approaches are a good start, but real disruptive changes spawn from a culture of—and investment in—innovation. Many organizations are simply not designed (intentionally or incidentally) to discover and invest in new opportunities. This is the crux of the innovator's dilemma: doing the right thing for the near and medium term is the wrong thing in the long run.

Many tried and true optimization techniques breed inflexibility into the system. M&As may get you a new tool or line of business, but they won't change how your organization sees the world. The antidote is to intentionally promote and invest in agility and innovation. Whether you choose to call this agile or digital transformation, the outcome should be the same: a counterweight to the inherent inflexibility we've designed through optimization, with the concerted promotion of flexibility and a portfolio of experimentation. As the world changes and deep tech moves into high tech, the organizations structured to take advantage of that change will thrive.

Finally, what if you're simply deciding which deep tech your company should invest in? As Nelson Mandela said, "Education is the most powerful weapon which you can use to change the world." Don't necessarily take others' word for it, and don't put all of your eggs in a single basket. In 1903, the esteemed president of the Michigan Savings Bank said, "The horse is here to stay, but the automobile is only a novelty—a fad." The fact is, finding quotes from respectable people who have wrongly predicted technology failures is easy—almost as easy as finding quotes from experts falsely predicting success. The future is unknown and unknowable, and innovation is a bit like gambling. A lack of certainty shouldn't unnerve anyone who is comfortable with making investment decisions, however. It just may require you to think a bit differently about risk and reward.

As we discussed in Chapter 1, *deep tech* refers to the stage the technology is in: impossible yesterday, barely feasible today, and soon to be so pervasive it's hard to remember life without it. Deep tech is an early phase of a general purpose technology, and deep tech solutions are reimaginations of fundamental capabilities. Deep tech is primed for investment: you place several smaller, early stage bets just before the technology is ready for mass market. The timing can be tricky, but some companies have realized that if they bet big and push hard enough, they can accelerate the timeline by years. Think Alphabet funding Waymo autonomous vehicles, or Facebook funding Oculus VR. Unless you're a huge enterprise deep with talent and funding, it can be hard to push an emerging technology into the market. But that doesn't mean you can't be aware of what the big players are doing and join them for the ride. AVs are floating many new industries, such as LIDAR tech, or the very concept of ride sharing.

In Chapter 1, we looked at this chart, which is a rough estimate of the financial possibilities around these seven deep technologies. Error bars being what they are, we're looking down the barrel somewhere between $50 trillion and $250 trillion in new global value creation over the next decade, thanks to deep tech.

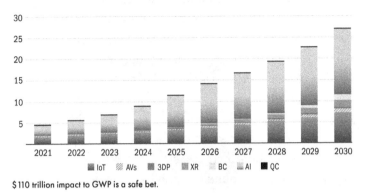

Deep Tech, Global Economic Impact (addition to GWP)

| | IoT | AVs | 3DP | XR | BC | AI | QC |

$110 trillion impact to GWP is a safe bet.

But the future is unknowable, and the numbers may even go beyond our generous ranges. Or the overall values may be correct, but the mix of technologies that end up as breakouts and the ones that lag behind are different than predicted. As with everything concerning large numbers, we tend to be more accurate in the aggregate. Will XR really be worth more than blockchain? That's anybody's guess. The wise move to make when faced with uncertainty is diversification. Divide your bets across several technologies, and you'll be in a good position to win.

AI. XR. Blockchain. IoT. AV. 3D Printing. Quantum computing. Any or all of these will change the decade and century. Whether you're an entrepreneur, business executive, or investor, there are plenty of roles to play to ensure that you, your tribe, and the world are better off with their existence. So lean into this once-in-a-century opportunity. People who are afraid to rock the boat rarely make the history books.

WHAT NEXT?

At the end of each chapter was a list of additional reading

material focused on each individual technology. It's only fair to wrap up this book with some writings that present a wider view beyond the individual technologies but discuss how they will come together to create something new. Check out the Further Reading sections at the end of each chapter.

However, self-education is only half of it. One of the realities of being an early adopter is that your world starts off pretty lonely. Almost by definition, there are relatively few folks to share your new ideas with. I've found two good ways of dealing with this reality. Optimally, you should apply both.

First, seek out those rare people in the world who share your passion. It's not always obvious up front, but a few hours spent on Google and in community forums can usually lead you to who you're looking for. One of the good outcomes of the balkanization of culture into smaller subgenres of tighter communities is that there is nothing too small or bizarre for a few dedicated fans around the world. Are you a forty-year-old male who is also a fan of the children's cartoon *My Little Pony*? Congratulations, you're a *bronie* and there's a community for you. When I first caught wind of this AR thing called Google Glass, I hopped on a plane to San Francisco to get one. Then founded a wearables Meetup group in Portland. Then I wrote two books about Glass and taught some classes.

Which brings us to point number two: educate others. Be part of the growth. Talent is always scarce in the realm of deep tech. This means that education must be a major component of any new technology revolution. The Winklevoss twins didn't merely invest in Bitcoin—they jet-set around the world, educating the financial establishment of its promise. They became ambassadors. They let their excitement infect others. When

Palmer Luckey wanted to experience VR, he didn't merely build his own headset and plug in, and bask in worlds of his own making. He wanted to share the creation with a community of VR enthusiasts. So he created a crowdfunding campaign that anyone could take part in. His goal was not to sell his VR company to Facebook but to bring VR to the world through access and education. Like most things in life, deep tech is more fun with friends.

MAGIC

"Is magic real?" Asked my wide-eyed three-year-old daughter, following an umpteenth viewing of Disney's movie *Frozen*. It's a common childhood question, so I should have been better prepared. My wife and I had long decided to explain the simple truth whenever possible.

"Is Santa real?"

"No, but it's still fun to pretend."

"Why is the sky blue?"

"The atmosphere scatters certain colors."

"Where do babies come from?"

Sigh. Knowing that children unearth wonder from the mundane anyway, we didn't see the point in heaping up untruths. Besides, as Mark Twain said, "If you always tell the truth, you don't have to remember anything." So, I answered honestly. "Yes."

How do I know magic is real? Because as Arthur C. Clarke

said, "Sufficiently advanced technology is indistinguishable from magic." I've saved an alien base from an army of malevolent robots from the comfort of a Las Vegas warehouse while wearing a VR headset. I've seen limbless children gain new 3D printed legs from a magic box. I've traveled the city without any human involvement via mobile app to AV. I've shaken hands with someone across the room and felt every sensation with the aid of haptic robot arms. I've hacked the very nature of reality to spill impossible answers just by uttering the right incantation using a quantum computer. None of these are experiments: they're all in production today. To paraphrase William Gibson, magic is real—it's just unevenly distributed.

In the last book of Marcel Proust's magnum opus, he returned to his childhood home years later and much wiser. The ravages of time were apparent, and his new-aged perspective diminished the giants of his youth. The problem with magic is that it's wild, and it acts with a will of its own. Only when we peer into the wild does magic lose its mysticism, and only then does magic become craft. The purpose of this book was to demystify deep tech and to reduce the giants into a manageable form. Now we can wield these tools properly and return to our homes to find that we are the new generation in charge. Others, seeing our prodigious outputs will ask of us, "Is magic real?" And we can say, "Yes."

FURTHER READING

- *The Future Is Faster Than You Think: How Converging Technologies Are Transforming Business, Industries, and Our Lives.* Peter H. Diamandis and Steven Kotler
- *The Inevitable: Understanding the 12 Technological Forces That Will Shape Our Future.* Kevin Kelly

- *Soonish: Ten Emerging Technologies That'll Improve and/or Ruin Everything.* Kelly Weinersmith and Zach Weinersmith
- *Machines of Loving Grace: The Quest for Common Ground Between Humans and Robots.* John Markoff
- *The Innovator's Dilemma: When New Technologies Cause Great Firms to Fail.* Clayton M. Christensen
- *The Third Wave: An Entrepreneur's Vision of the Future.* Steve Case
- *Machine, Platform, Crowd: Harnessing Our Digital Future.* Erik Bryjolfsson and Andrew McAfee
- *Mapping Innovation: A Playbook for Navigating a Disruptive Age.* Greg Satell
- *Crossing the Chasm: Marketing and Selling Technology Projects to Mainstream Customers.* Geoffrey A. Moore

ACKNOWLEDGMENTS

It's a cliché to say, but like all books, this one is a team effort with more people involved to name, but I'll embarrass a few. Thanks to my innovation team, who read far too many unpolished words and too much half-thought prose: Heidi, Shashwat, Vic, Roger, Colin, Ron, Thyanna. Thanks to the folks in endless research labs and startups who let me peek into the future: MIT Media Lab, MIT Bootcamp, Microsoft Research Lab, Google X, World Economic Forum, Xerox PARC, Waymo, HaptX, MetaVRse, Senrio, and so many more—you know who you are. Thanks to the editors, designers, layout artists, marketers, managers, and others who turned a word cloud into a finished work. Finally, thanks to my wife, Noelle, a voice of reason and the perfect test audience: smart, non-techy, and brutal.

ABOUT THE AUTHOR

ERIC REDMOND is the Forrest Gump of technology: a twenty-year veteran technologist who always happens to show up wherever deep tech history is being made, from the first iPhone apps to big data to Bitcoin. He has advised state and national governments, Fortune 100 companies, and groups as varied as the World Economic Forum and MIT Media Lab. He has also authored half a dozen technology books (including two tech books for babies) and spoken on every continent except Antarctica. Today, he's a husband, a dad, and the leader of a global tech innovation team.

CPSIA information can be obtained
at www.ICGtesting.com
Printed in the USA
LVHW032038030722
722682LV00003B/299

9 781544 518947